Artificial Intelligence and Internet of Things for Smart Agriculture

The purpose of this book is to explore the fundamentals and implementation strategies of smart agriculture through the integration of artificial intelligence and Internet of Things technologies. It examines critical aspects of planning and decision-making while providing recommendations for implementing efficient and effective intelligent agriculture solutions. Readers will gain valuable insights into how these smart technologies enhance productivity, sustainability, and precision in contemporary technology-driven farming systems.

This book:

- Discusses artificial intelligence-based 3D mapping of soil, smart weeding systems, and crop visualization in smart agriculture.
- Covers data analytics-based decision-making, improving crop quality based on precise prediction, minimizing the use of pesticides for better soil quality, and precision farming systems.
- Examines problems like weather forecasting, crop prediction, calculation of pesticide quantity, irrigation requirement, and prediction of the right time for seeding.
- Highlights the role and importance of data visualization techniques in smart agriculture implementation.
- Presents the internet of things and artificial intelligence-based solutions for smart agriculture.

It is primarily written for senior undergraduate, graduate students, and academic researchers in the fields of electrical engineering, electronics, and communications engineering, computer science and engineering, and agricultural science and engineering.

Artificial Intelligence and Internet of Things for Smart Agriculture

Edited by
Praveen Kumar Shukla, Harshal A. Sanghvi,
Ankur Agarwal, and Abhijit S. Pandya

CRC Press
Taylor & Francis Group
Boca Raton London New York

CRC Press is an imprint of the
Taylor & Francis Group, an **informa** business

Designed cover image: Shutterstock

First edition published 2026
by CRC Press
2385 NW Executive Center Drive, Suite 320, Boca Raton FL 33431

and by CRC Press
4 Park Square, Milton Park, Abingdon, Oxon, OX14 4RN

CRC Press is an imprint of Taylor & Francis Group, LLC

ISBN: 9781032565873 (hbk)
ISBN: 9781041147794 (pbk)
ISBN: 9781003676102 (ebk)

DOI: 10.1201/9781003676102

Typeset in Sabon
by Newgen Publishing UK

Contents

Foreword

Agriculture, one of the world's oldest and most vital professions, has long been responsible for sustaining the population. Today, the integration of advanced technologies is transforming agriculture, significantly improving both the quality and quantity of production. Among the most promising innovations are Artificial Intelligence (AI) and the Internet of Things (IoT), which are driving the successful adoption of smart agriculture. These technologies are helping to build an intelligent ecosystem that enhances farming efficiency.

IoT devices, equipped with sensors, collect crucial data—such as soil health, weather conditions, and crop status—and process it for informed decision-making and strategic planning. This enables key agricultural functions such as data-driven farming, optimization of agricultural practices, and the monitoring of livestock, irrigation systems, and supply chains. AI-powered IoT systems are further revolutionizing modern farming by enabling precision techniques like smart irrigation, livestock management, and supply chain optimization.

This book delves into these transformative processes, emphasizing precision agriculture and automation. It explores key topics such as food quality assurance, weather prediction, 3D soil mapping, pesticide application through drones, robotics for labor management, and innovative technologies like chatbots and smart weeding systems. With its focus on these cutting-edge tools, I am confident that this book will serve as a valuable resource for those looking to implement smart agriculture.

I extend my best wishes for the success of this book.

Vrijendra Singh
Indian Institute of Information Technology Prayagraj

Preface

Agriculture remains at the heart of human survival, delivering food and resources to sustain the growing global population. Considering increasing challenges such as climate change, population growth, and resource constraints, the need for innovation in agriculture is a prime focus. Traditional farming methods are giving way to smarter, more efficient practices, driven by the integration of advanced technologies.

This book focuses on **Artificial Intelligence (AI)** and the **Internet of Things (IoT)**—two cutting-edge technologies that are transforming agriculture and laying the foundation for **smart farming**. These innovations are helping farmers make data-driven decisions, optimize their agricultural practices, and increasing the quality and quantity of their produce. By utilizing IoT devices for real-time data collection and AI for advanced processing and decision-making, we are witnessing a new era in which technology meets tradition for more sustainable and productive farming.

Chapter 1 introduces the applications of AI and drones in smart agriculture, focusing on precision and automation. In Chapter 2, the enhancement of food quality assurance through IoT is explored. Chapter 3 delves into the use of machine-learning techniques for weather prediction. Chapter 4 gives an in-depth look at the use of chatbots in agriculture. Chapter 5 covers data-driven smart agriculture, emphasizing the role of machine learning, while Chapter 6 discusses 3D soil mapping

The challenges and opportunities of drone-assisted pesticide spraying are the focus of Chapter 7, with IoT-based smart irrigation systems explained in Chapter 8. Chapter 9 tackles the use of robotics to address labor shortages in agriculture, while Chapter 10 revisits smart irrigation, highlighting machine learning applications. The impact of drones on agriculture is revisited in Chapter 11, and Chapter 12 explores livestock management using IoT. An examination of AI and IoT in the agricultural supply chain is mentioned in Chapter 13. Finally, Chapter 14 introduces smart weeding systems, completing the exploration of advanced technologies in modern agriculture.

This book is intended for researchers, practitioners, and students interested in the intersection of agriculture and technology. It offers not only theoretical insights but also practical solutions to real-world challenges, making it an essential guide for anyone involved in or studying **smart agriculture**.

As we stand at the forefront of a technological revolution in farming, we hope this book serves as a comprehensive resource that will inspire further research and innovation in this exciting and rapidly evolving field.

Praveen Kumar Shukla
Harshal A. Sanghvi
Ankur Agarwal
Abhijit Pandya

Editor Biographies

Praveen Kumar Shukla is currently serving as a Professor and Dean at the School of Engineering, Babu Banarasi Das University, Lucknow, UP, India. He is also an Adjunct Faculty at the SAU Centre for Research and Innovative Learning (SCRIL), South Asian University, New Delhi. Dr. Shukla is also the executive committee member and Secretary of IEEE CIS UP Chapter. Dr. Shukla holds a Ph.D. in Computer Science & Engineering from Dr. A.P.J. Abdul Kalam Technical University, Lucknow. He also holds M.Tech in Computer Science & Engineering and B.Tech. in Information Technology. His research area includes Fuzzy Systems (Interval Type-2 Fuzzy Systems and Type-2 Fuzzy Systems), Machine Learning, Computer Vision and Deep Learning. He has published more than 60 Papers in reputed International Conferences and International Journals. He has been honored as keynote speaker and session chair at national/international conferences, faculty development programs and workshops.

Harshal A. Sanghvi is a Visiting Instructor and Director of Healthcare Analytics at the College of Business, Florida Atlantic University. Dr. Sanghvi has a Ph.D. in Computer Engineering from Florida Atlantic University. He has been teaching and training professionals in areas of Artificial Intelligence and Machine Learning, Immersive Learning, Data Mining, Business Analytics, Social Media, and Web Analytics. His research is focused on Engineering, Applied Sciences, Healthcare Analytics, Computer Vision, Clinical Decision Support Systems, Data Analytics, Neural Networks, Medical Imaging, the Internet of Things (IoT), and Artificial Intelligence. He has a blend of industrial and academic experience, which allows him to teach and train professionals with industrial use-cases.

Ankur Agarwal is a Professor at Florida Atlantic University, USA. He has over 15 years of Academic, Administrative and Industry Experience. He has more than 80 publications in peer reviewed national and international journals, conferences, and book chapters. He has several national and international patents. He has been awarded grants by the National Science Foundation and National Institute of Health and has served as a PI and Co-PI in several projects.

Abhijit S. Pandya is a Professor at Florida Atlantic University, USA and also a Director of Clinical Trials and Research at Specialty Retina Center, FL, USA. He has over 33 years of Academic, Administrative and Industry Experience. He has more than 100 publications in peer reviewed national and international journals, conferences, and book chapters. He has several national and international patents. He has been awarded grants by the National Science Foundation and National Institute of Health and has served as a PI and Co-PI in several projects.

Contributors

Alam, Mohammad Nazmul
Faculty of Computing, Guru Kashi University, Talwandi Sabo, Bathinda, Punjab, India

Ahmad, Faiyaz
Department of Computer Science and Engineering, Integral University, Lucknow, India

Agrawal, Ankit
Hindustan Aeronautics Limited, Lucknow, India

Agarwal, Rupali
Department of Electronics and Communication Engineering, Babu Banarasi Das Institute of Technology and Management, Lucknow, India

Agarwal, Vedaant
Pine Crest, Fort Lauderdale, Florida

Anusha, P.
Department of Computer Science and Engineering, Guru Nanak Institute of Technology, Hyderabad, Andhra Pradesh, India.

Auroshaa, A.
Department of Information Technology, Hindustan Institute of Technology and Science, Padur, Chennai, India

Babu, C. V. Suresh
Department of Information Technology, Hindustan Institute of Technology and Science, Padur, Chennai, India

Das, Sanjai A
Department of Information Technology, Hindustan Institute of Technology and Science, Padur, Chennai, India

Dubey, Vandana
C. Byregowda Institute of Technology, Kolar, Karnataka, India

Ghai, Rajesh Kumar
Faculty of Management and Commerce, Swami Vivekanand Subharti University, Meerut, India

Ghosh, Partha
Assistant Professor, Department of Computer Sc. and Engineering, Govt. College of Engineering and Ceramic Technology, Kolkata

Ghosh, Suradhuni
Assistant Professor, Department of Geography, Govt. Girls' General Degree College, Ekbalpur, Kolkata

Hymavathi, M
Dept of Computer Science and Engineering, N.B.K.R.I.S.T, Vidyanagar, Andhra Pradesh, India

Jahnavi, Y.
Dept of Computer Science, Government Degree College, Naidupet, Andhra Pradesh, India.

Joshi, Saket
Department of Electronics and Instrumentation, School of Technology, Nirma University, Ahmedabad, Gujrat, India

Kamal, Anum
Department of Computer Science and Engineering, Integral University, Lucknow, India

Kamal, Aisha
Department of Bioengineering, Integral University, Lucknow, India

Kumar, Sachin
Parul University, Baroda, Gujarat, India

Kumar, P. Nagendra
Department of Computer Science and Engineering, Geethanjali Institute of Science and Technology, Nellore, Andhra Pradesh, India

Kumari, Priti
Ashoka Institute of Technology and Management, Varanasi, U.P., India

Parikh, Riya
IIT Kharagpur

Patel, Kavita
Ashoka Institute of Technology and Management, Varanasi, U.P., India

Patel, Swati
Acharya Motibhai Patel Institute of Computer Studies
Ganpat University, Mehsana, India

Parikh, Satyen
Acharya Motibhai Patel Institute of Computer Studies
Ganpat University, Mehsana, India

Patel, Savan
Acharya Motibhai Patel Institute of Computer Studies
Ganpat University, Mehsana, India

Patel, Haresh
Matter Motor Works Pvt. Ltd
Ahmedabad, India

Rani, Aarti
Babu Banrarsi Das University, Lucknow, India

Singh, Nishu
San Diego State University, CA

Singh, Vijendra Pratap
Mahatma Gandhi Kashi Vidyapeeth, Varanasi, U.P., India

SandeepuKumar, K.
Dept of Computer Science, Gayatri Vidya Parishad College for Degree and P.G Courses (A), Rushikonda, Visakhapatnam, Andhra Pradesh, India

Singh, Gurpreet
Faculty of Computing, Guru Kashi University, Talwandi Sabo, Bathinda, Punjab, India

Singh, Sukhpreet
Faculty of Computing, Guru Kashi University, Talwandi Sabo, Bathinda, Punjab, India

Sharma, Gaurab Kumar
Faculty of Management and Commerce, Swami Vivekanand Subharti University, Meerut, India

Yamini, B.
Dept of Networking and Communications, School of Computing, College of Engineering and Technology, SRM Institute of Science and Technology (SRMIST), SRM Nagar, Kattankulathur, Chengalpattu District, Tamil Nadu, India

Chapter 1

Artificial Intelligence and Drones

Revolutionizing Smart Farming through Precision and Automation

Anum Kamal, Faiyaz Ahmad, and Aisha Kamal

1.1 INTRODUCTION

The agricultural sector, comprising crops, forestry, livestock, and aquaculture, plays a crucial role in global sustenance [1], facing escalating pressures due to rapid population growth. It is predicted that there will be 10 billion people on the planet by 2050 [2], driving a 70% surge in food demand, as predicted by the Food and Agriculture Organisation of the United Nations (FAO) [3]. This places immense pressure on food production systems, exacerbating existing challenges like soil testing, weed management, and limited irrigation access [4, 5]. Conventional farming methods contribute to environmental degradation and food insecurity due to issues like pesticide-resistant pests and health hazards from chemical use [6].

To address urgent agricultural concerns, a shift towards sustainability is imperative, facilitated by digital technologies such as the IoT, AI, and cloud computing [7]. The goal is to attain a 70% productivity rise by 2050, with capital in the farming industry having surged by 80% over the previous five years [8, 9]. These technologies hold significant potential for enhancing productivity sustainably, crucial for meeting the growing demands of the global population amidst dwindling cultivable land [10].

This chapter explores AI and drone roles in crop monitoring, mapping, spraying, livestock oversight, soil analysis, autonomous farming, and decision support systems. It emphasizes the fusion's advantages, like timely disease detection, resource optimization, reduced environmental impact, and increased profitability. The narrative extends to future prospects and potential evolution, contributing to a comprehensive understanding of profound impact of AI and drones on modern agriculture. Figure 1.1, depicts the evolution and advancements in the field of agriculture.

DOI: 10.1201/9781003676102-1

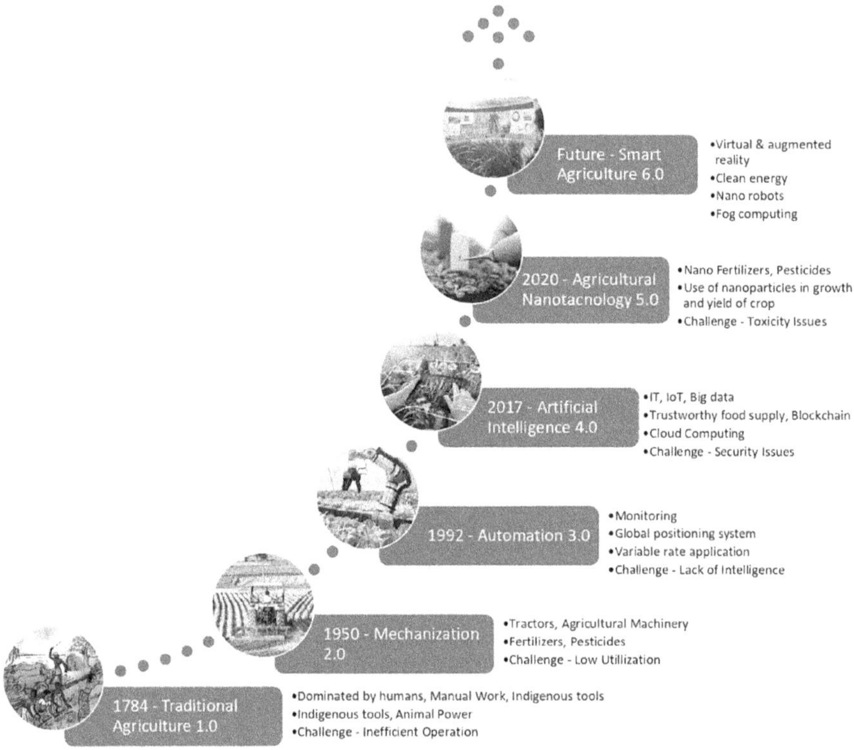

Figure 1.1 Roadmap of Agricultural Development.

1.2 MAJOR OBSTACLES IN AGRICULTURE

Farmers contend with a diverse range of challenges (Figure 1.2), several of which can be effectively tackled through the implementation of precision agriculture.

According to the FAO, plant diseases represent a major threat to agriculture and cause large annual losses in global GDP of over 220 billion USD [11]. Farmers rely on visual inspection of fields to detect signs of infection, followed by laboratory examination of plant samples for accurate diagnosis, enabling timely implementation of corrective measures to mitigate spread and yield losses [12].

Urban vegetation classification is another common challenge that is pivotal for addressing climate change. Machine learning aids in identifying tree species diversity. Integrating spectral data from Unmanned Aerial Vehicle

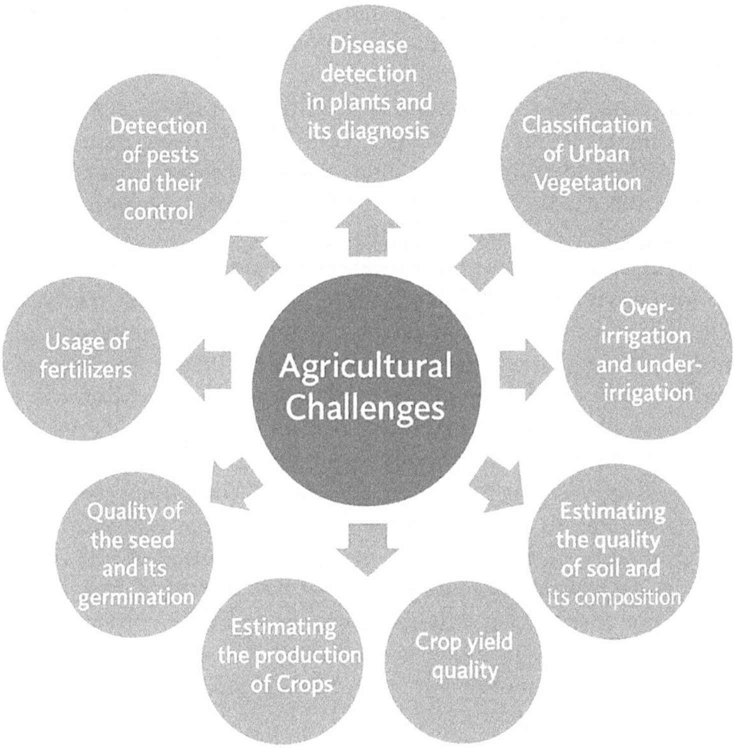

Figure 1.2 Agricultural Challenges.

(UAV) images enhances accuracy in tree selection, crucial for understanding ecosystem dynamics and carbon storage processes affected by dominant species, and combating climate challenges [13].

The third challenge for farmers is managing irrigation which is critical for postharvest crop quality [14]. Under-irrigation significantly reduces yields [15], while over-irrigation impacts yield and fruit quality [16]. UAV multi-spectral imagery aids in water management by detecting water levels, crucial for addressing water scarcity [17]

Furthermore, soil quality is vital for crop output and yield potential [14]. UAVs with multispectral cameras capture geospatial data, including water and nutrient levels. This aids in identifying soil attributes requiring attention and facilitates timely soil treatments, such as foliar sprays, to promote optimal crop development [18].

Moreover, timely harvesting and tailored storage plans are crucial for maintaining crop quality post-harvest. Assessments like measuring color, size, firmness, and acidity are employed to uphold these qualities.

Post-harvest, farmers devise storage plans tailored to factors like duration, temperature, and humidity, using methods such as Dynamic Controlled Atmosphere (DCA) storage to ensure optimal conditions for preserving crop quality [14].

Accurate crop yield estimation is essential for labor deployment and storage strategies [19]. Traditional methods are inaccurate and labor-intensive. Advanced technologies offer efficient solutions, enabling precise yield estimation across diverse fields, and enhancing decision-making and resource allocation in agriculture [20].

Seed quality and viability are crucial for crop yield and quality. Traditional monitoring methods are labor-intensive, but UAV-based methods offer efficient and effective seed and seedling monitoring, revolutionizing agriculture by providing real-time supervision [21]. Proper fertilizer usage is essential for crop yield, as inadequate or excessive fertilization can impact crop quality. UAV technology enables precise monitoring and application of fertilizers, ensuring optimal crop nutrition and quality [22].

A balanced approach to fertilization is crucial for optimal crop outcomes. Utilizing UAVs allows precise and controlled fertilizer applications, enhancing crop quality. Detecting and controlling pests poses a significant challenge in agriculture, potentially leading to substantial crop yield losses [23]. Traditional detection is labor-intensive and time-consuming, relying on expert input.

Pesticide use is common for pest management, but it comes with adverse health consequences and fosters the development of resistant pest populations. Over 500 insect species and around 270 weed species are already resistant [24]. The sustainability of pesticide application is questionable amid mounting resistance and environmental concerns. Innovative and sustainable pest control methods are needed to address these challenges and ensure effective pest management as agriculture advances.

1.3 SMART FARMING AND PRECISION AGRICULTURE

Smart farming utilizes information technologies to optimize agricultural systems, integrating information and communication technologies to enhance overall productivity [1]. At its core, AI plays a pivotal role, with applications ranging from intelligent machines to sensors that replicate human actions. AI integration with traditional agriculture leads to smart agriculture, incorporating techniques like precision farming for improved crop monitoring and management, ultimately enhancing economies [10].

A range of monitoring tools and sensors are used in precision agriculture to evaluate farming factors such as soil health, nutrient levels, water content, and vegetation greenness [25]. These measurements aid farmers in making informed decisions, reducing resource wastage, and increasing yield [26]. It saves time and reduces manual labor, with Unmanned Ground Vehicles (UGVs) providing a cost-effective alternative to UAVs, albeit with speed and coverage limitations. UAVs, emerging as a promising alternative to satellites and UGVs, offer a cost-effective and versatile solution [27].

1.4 ARTIFICIAL INTELLIGENCE AND AGRICULTURE

Artificial Intelligence (AI) replicates human intelligence in digital devices, computer systems, and robotics [28]. AI's foundation lies in three cognitive skills: learning, reasoning, and self-correction [29]. It entails creating computer systems that can do tasks related to human intelligence, such as prediction, pattern recognition, and experiential learning [30]. AI is also a game-changer in the field of smart farming (Figure 1.3).

Smart sensors, computer vision, image processing, expert systems and machine learning address agricultural challenges, enhancing quality, productivity, and sustainability. In precision agriculture, AI algorithms optimize cultivation strategies using data from weather stations and sensors [30]. AI aids in soil testing, pest control, disease treatment, irrigation, and post-harvest operations, driving agricultural innovation [31, 32, and 33].

1.4.1 The Basics of AI Technologies Implemented in Agriculture

Initial AI integration involves machine learning, processing input data to extract relevant information from specific databases. Real-time data interpretation is crucial for AI's conclusions, often guided by weather forecasts for crop season decisions. UAVs aid threat detection like forest fires and pest infestations through geolocation [32]. Problem-solving methods include expert systems, Artificial Neural Networks (ANNs), neuro-fuzzy logic, and fuzzy logic with ANNs leading, mimicking human brain processes [33].

1.4.2 Use of AI in Selection of Crop or Seed

Successful agricultural growth relies on robust vitality, germination rate, and seedling emergence, ensuring uniform production [34]. AI-driven tools like SeedGerm utilize historical data to comprehend seed responses, outperforming traditional methods. This platform employs automated seed imaging and machine learning for germination scoring and morphological analysis, facilitating seed batch performance evaluation and issuing germination certificates [35]. Additionally, the Crop Selection Method (CSM) introduced by Kumar et al. (2015) offers a novel technique for crop selection [36].

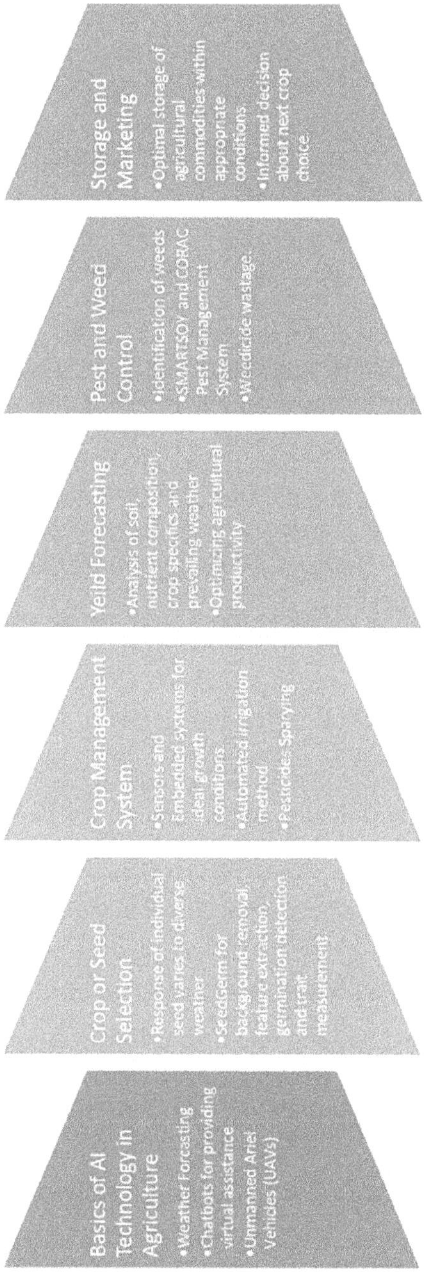

Figure 1.3 Role of Artificial Intelligence in Agriculture.

1.4.3 AI-driven Crop Management System

Advanced technologies like sensors and embedded systems optimize crop growth conditions by regulating factors such as light intensity and humidity in real time [37]. Trace Genomics utilizes DNA analysis and machine learning to provide evidence-based insights for crop selection and land suitability. AI aids in real-time monitoring of soil moisture for efficient water management, while innovations like the 'Rover' device reduce pesticide usage and enhance crop coverage [38].

1.4.4 AI-driven Yield Forecasting

Among the array of AI techniques, prediction models have emerged as a cornerstone readily embraced by farmers. These models delve into an intricate analysis encompassing soil characteristics, nutrient composition, crop specifics, and prevailing weather conditions as integral components to forecast future yield outcomes [39]. This emphasizes the pivotal role prediction models play in shaping informed decision-making and optimizing agricultural productivity.

1.4.5 AI-driven Pest and Weed Control

Partel et al. (2019) [40] present an inventive approach, detailing the development of automated machines designed for the precise identification of weeds, subsequently administering targeted applications of agrochemicals. This innovation not only curtails weedicide wastage but also mitigates crop exposure to agrochemicals. Notable instances of pest management systems that embody such innovative advancements include the SMARTSOY and CORAC projects [5]. These pioneering endeavors showcase the potential of technology-driven solutions to revolutionize pest management practices and enhance agricultural sustainability.

1.4.6 AI-enhanced Product Storage and Marketing

Maintaining proper storage conditions for agricultural commodities is crucial to preserve product quality. Innovative sensor arrays integrated into storage chambers extend post-harvest shelf life, revolutionizing traditional farming practices. Advanced technologies (Figure 1.3) now provide detailed market analysis, guiding strategic decisions on crop choices [32]. This integration of technology and market insights marks a paradigm shift in agriculture, enhancing efficiency and decision-making quality in the industry.

1.5 DRONE TECHNOLOGY IN FARMING

Over the past three and a half decades, remote sensing has emerged as an indispensable cornerstone of Precision Agriculture and Smart Farming, revolutionizing the way cultivated fields are monitored and managed. This technology has been instrumental in providing sophisticated solutions to enhance agricultural efficiency and productivity, marking a significant milestone in the evolution of modern farming practices [41]. It employs images captured across different wavelengths to monitor various crops and vegetation. Remote sensing methods, previously reliant on satellite imagery or manned aircraft, focused on specific growth stages to evaluate vegetation status. However, challenges like compromised spatial resolution, limited temporal availability, and environmental interference with satellite imagery, such as cloud cover, hindered their effectiveness. Manned aircraft deployment incurred high costs and logistical constraints, restricting the capture of only a few crop images. The emergence of UAV-based remote sensing systems marks a transformative leap forward, propelling remote sensing and Precision Agriculture to new heights. UAVs revolutionize crop monitoring with swift, cost-effective data acquisition. Integrated with IoT technology, they offer high-resolution imaging, enhancing Precision Agriculture's remote sensing capabilities. Operating at low altitudes, UAVs provide flexibility in image capture, outperforming manned aircraft and ground systems. Their efficiency and simplicity unlock diverse agricultural applications, promising significant benefits in monitoring and management [42].

1.5.1 Drone System Technology

Unmanned Aerial Systems (UAS) for Precision Agriculture consist of Ground Control Stations (GCS), Unmanned Aerial Vehicles (UAVs), UAV Control Systems (UAV CS), and sensors for data acquisition. UAVs perform various tasks autonomously or via remote control. GCS manages communication and control, providing real-time flight and sensor data. UAV CS oversees flight operations and processes sensor data. Sensors capture crop, soil, and environmental data crucial for decision-making. In tasks like pesticide spraying, sensors can be replaced with task-specific components. Integration of these elements enables UAS to conduct diverse Precision Agriculture tasks, enhancing crop management and resource utilization [43].

1.5.2 Design-based Categorization of UAV Technology

When considering UAVs used for Precision Agriculture, these technologies can be categorized into five fundamental categories:

- *Fixed-wing*: These are fixed-wing UAVs resemble unmanned planes, requiring runways or catapults for takeoff. They excel in covering vast areas due to high endurance and speed but tend to be costlier. Solar-powered variants promise extended flight times [44].
- *Rotary-wing*: These are also referred to as rotorcraft or Vertical Take-Off and Landing (VTOL) UAVs. They provide stable hovering and maneuverability, although they lack high speeds and long flight times. Widely utilized in agriculture, they are cost-effective for monitoring smaller crop areas. Rotary-wing UAVs include:
  - Unmanned helicopters: Equipped with main and tail rotors, mimicking conventional helicopters
  - Multi-rotor: UAVs with four or more rotors (quadcopter, hexacopter, octocopter, and the like.) that provide stability and maneuverability.
- *Blimps*: Lighter-than-air UAVs boast high endurance and larger sizes, flying at low speeds. They can stay afloat without power and are safer in collisions. Despite these advantages, they are less prevalent in Precision Agriculture.
- *Flapping wing*: Small UAVs with flexible wings mimic bird and insect flight but have high energy consumption, making them less common in Precision Agriculture.
- *Parafoil-wing*: With propellers at the rear, these UAVs control flight direction efficiently, minimizing energy use. Although they carry larger payloads, they are less prevalent in Precision Agriculture.

Multi-rotor UAVs are commonly used for agricultural tasks due to their ease of operation, slower speeds, maneuverability, and cost-effectiveness. These advantages facilitate imaging and data collection from crops, which aligns with the main application of UAVs in vegetation monitoring. However, fixed-wing UAVs are preferred for monitoring larger areas efficiently.

1.5.3 Sensors Used for the Acquisition of Data in UAVs

UAVs, with specialized sensors, augment IoT techniques in agriculture. These sensors capture high-resolution images vital for monitoring vegetation characteristics. Sensor selection for agricultural UAVs hinges on crop parameters, with emphasis on low weight, energy consumption, small size, and high-resolution image capture [9]. Prominent sensor types for Precision Agriculture include visible light, multispectral, hyperspectral, and thermal sensors. Visible light sensors capture the color and texture details of vegetation, while multispectral sensors analyze beyond human vision for insights into plant health. Hyperspectral sensors offer detailed crop analysis, and thermal sensors detect temperature variations and stress levels. Other sensors like Light Detection and Ranging (LiDAR), or laser scanners, find use in agriculture despite higher costs. Although well-established in environmental sciences, their adoption in UAV-based agricultural studies remains limited [45]. Various sensors monitor vegetation aspects like color, texture, shape, and radiation levels. Processed data aids in tracking crop traits such as biomass, health, and soil moisture, empowering UAVs in Precision Agriculture for informed decision-making.

1.5.4 Types of Data Acquired Using UAVs

UAVs or drones, are versatile tools capable of gathering various types of data across numerous applications. They can capture high-resolution visual images through photography and video footage, providing detailed information about landscapes and structures. Multispectral and hyperspectral cameras enable crop health analysis by detecting variations in vegetation health and growth. Thermal cameras detect temperature variations useful for thermal mapping and identifying heat leaks. LiDAR-equipped UAVs create accurate 3D models for topographic mapping and land surveying. Aerial surveying and mapping involve generating orthomosaic maps and Digital Elevation Models (DEMs) for terrain analysis.

UAVs play vital roles in agriculture, monitoring crop health, growth, and yield for precision farming. They assist in field management tasks like irrigation planning and pest control. The data collected by UAVs depends on their sensors and mission objectives, highlighting their adaptability and utility across diverse fields.

1.5.5 Data Preprocessing Techniques

The utilization of UAVs in precision agriculture facilitates the collection of extensive data, which can undergo diverse processing and analysis techniques to extract valuable insights. This wealth of information enables farmers to make informed decisions, optimize resource allocation, and enhance overall agricultural productivity.

- *Preprocessing of Images*: Photogrammetry in precision agriculture reconstructs scenes using overlapping images, creating 3D models from 2D data. Utilizing UAVs enhances resolution for detailed models. Orthomosaics and Digital Elevation Models (DEMs) are generated, offering insights into vegetation structures like height and density. These models, combined with RGB imagery, facilitate applications such as vegetation indices mapping. Photogrammetry is widely used in precision agriculture, providing crucial 3D information for various applications [46, 47].
- *Analysis of Multi-spectral Data*: Multispectral sensors on UAVs capture data across various spectral bands like NIR and thermal

infrared, enabling simultaneous imaging. Processed data yields maps and analyses, often involving calibration and alignment for accurate comparisons. Techniques like radiometric and atmospheric correction enhance data quality. The analysis assesses vegetation health, crop conditions, and environmental changes like deforestation and erosion. It aids in land classification, identifying land use patterns, and evaluating urban development impacts.

- *Calculation of vegetation indices*: Vegetation Indices (VIs) are crucial in Precision Agriculture remote sensing. They combine spectral bands to highlight vegetation traits, minimizing external influences for reliable crop monitoring. Multiple VIs are extracted to deduce various conclusions, computed from individual photos or comprehensive orthophotos. These indices reveal valuable crop characteristics, spanning biological and physical aspects of vegetation.
- *Machine Learning*: Machine Learning (ML) is pivotal in processing data for prediction and identification tasks, spanning diverse domains such as medical systems [48], marketing [49], and biology [50]. In this context, ML is pivotal in Precision Agriculture, leveraging data from UAVs for tasks like estimating crop growth rates, disease identification, and object recognition. ML's popularity has surged, driven by rapid advancements, especially in deep learning, making it a powerful tool for analyzing vast agricultural datasets.
- *Geographic Information System (GIS)*: GIS is integral in preprocessing data from UAVs, capturing, analyzing, and presenting spatial data to aid decision-making. Integrated with UAV data, GIS enhances information quality and applicability. It combines images from multiple flights, filters, enhances imagery, and conducts land cover classification, and spatial and elevation analysis. Preprocessed datasets from GIS can be exported for further analysis or stakeholder sharing.
- *3-dimensional Modelling*: Three-dimensional (3D) modelling of agricultural landscapes, derived from UAV-collected data with LiDAR or photogrammetry sensors, provides detailed elevation information for creating Digital Elevation Models (DEMs) and Digital Surface Models (DSMs). These models reveal topographical features, aid in volume calculations for structures like stockpiles, and enable plant counting. The line-of-sight analysis identifies obstructions, while soil fertility maps correlate elevation data with nutrient content.

To expedite the data processing process, various software tools and techniques have been developed. The most frequently employed software tools are outlined in Table 1.1.

Additionally, some other software tools such as Erdas Imagine [54], eCognition [55], and PixelWrench 2 [56], offer promising capabilities for enhancing the efficiency of data processing in precision agriculture.

Table 1.1 Most Frequently Employed Software Tools for UAV Data Preprocessing

S. No.	*Ref.*	*Tool*	*About*
1.	[51]	QGIS	Typically utilized to compute vegetation indices using multispectral data.
2.	[51]	MATLAB	Primarily employed for computing vegetation indices, but can also be used for various other image-processing models.
3.	[52]	Pix4D	This tool, frequently employed, is the most widespread. It has applications in calculating VIs and/or creating orthomosaics and 3D models.
4.	[53]	Adobe Photoshop	Utilized to rectify misrepresentation and employ alternative techniques in the realm of image processing.
5.	[53]	Agisoft Photoscan	Employed in the production of orthomosaics and three-dimensional models. Additionally, it enables the computation of vegetation indices.

1.5.6 UAV-assisted Crop Observation

UAVs are versatile tools for crop monitoring across various crops, including cereals, fruits, vegetables, oilseeds, and specialty crops like coffee, tea, cocoa, and tobacco, aiding growth stage assessment, yield prediction, disease detection, and harvest optimization in agriculture.

However, it is important to acknowledge that using UAVs for crop monitoring comes with limitations [56, 57] including restricted coverage area per flight, susceptibility to weather conditions, and the need for skilled operators and specialized equipment, alongside complex regulations, adding deployment challenges. Considering these limitations is crucial when integrating them into crop monitoring programs. Despite benefits like improved management and productivity, considerations like crop characteristics and weather conditions are essential for effective monitoring.

1.5.7 Application of Drones

UAVs in smart farming offer soil analysis, precise planting, efficient crop spraying of pesticides, herbicides, and fertilizers [58], irrigation management to optimize water consumption and minimize waste [59], yield estimation of crops and livestock monitoring [60], vegetation health monitoring [61], and weed mapping through Site-Specific Weed Management (SSWM) [62], enhancing productivity, reducing environmental impact, and optimizing resource use. Smart farming has been transformed by the integration of AI and cloud technologies with UAVs, which has allowed for precision agriculture, increased efficiency, reduced environmental effects, and enabled

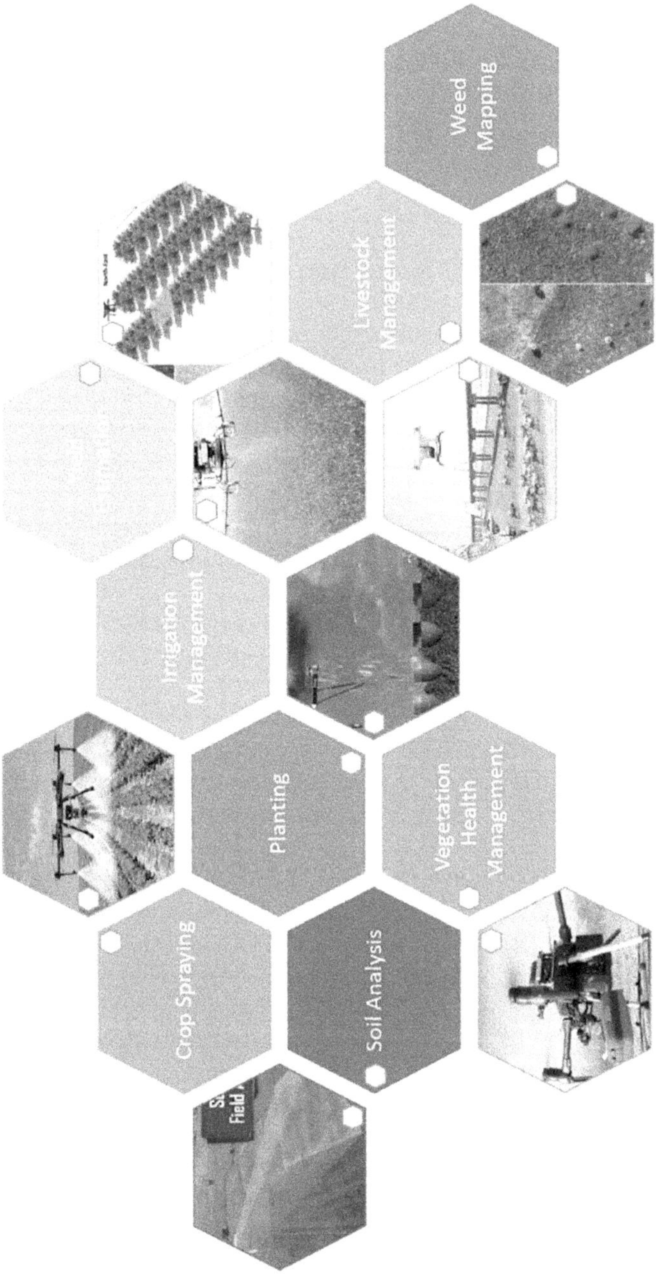

Figure 1.4 Applications of Drones.

real-time monitoring (Figure 1.4). UAVs are potent tools that benefit multiple industries and promote sustainability.

1.5.8 Challenges Associated with the Use of Drones in Agriculture

Although the application of UAVs in precision agriculture has shown great promise, there are currently several obstacles preventing their broader adoption:

- A lack of standardized workflows for UAV-based PA can deter adoption and lead to confusion and inefficiencies.
- PA's data-intensive nature requires skilled personnel for complex analysis, posing a barrier for resource-limited farmers.
- High investment costs for UAVs present a significant barrier to adoption, especially for smaller agricultural operations, as evidenced by data from the European Union (EU) indicating financial challenges for many smaller farmers [63].
- Limited flight time (20 min. to 1 hr.) in commercial UAVs restricts coverage area, requiring multiple flights for larger fields, with longer-flight options often costly.
- UAV flights are heavily influenced by weather conditions, with adverse conditions like strong winds or heavy rain hindering scheduling and data collection.
- Small-scale farming, common in regions like the EU, poses cost-effectiveness challenges for implementing UAV-based PA practices.
- Regulatory considerations for UAV use in agriculture, can still pose complexities, requiring adherence and authorizations.

EU regulations [64] categorize UAV operations into Open, Specific, and Certified categories based on risk levels. The Open category covers low-risk operations within Visual Line Of Sight (VLOS) and do not require special authorization or pilot licensing. The Specific category involves medium-risk operations requiring risk assessment and mitigation. The Certified category, aligned with manned aviation rules, pertains to high-risk operations in controlled airspaces, not directly related to Precision Agriculture.

In summary, UAVs hold promise for Precision Agriculture, but challenges like standardized workflows, training, cost, flight times, weather resilience, and regulations must be addressed for wider adoption. Anticipated advancements in technology and sensor utilization could reduce costs and overcome limitations like short flight times, making UAVs more accessible and effective tools for farmers in the future.

1.6 THE DIFFICULTIES AND SUGGESTIONS REGARDING THE INTEGRATION OF AI IN MODERN AGRICULTURE

As the global population increases, technology integration in agriculture becomes essential for higher-quality food production. While some farmers benefit from modern technology, many struggle to improve yields. Developed countries like Australia and parts of Europe have seen significant yield growth through advanced technology adoption, emphasizing its positive impact on yield and sustainability [65].

Agriculture is evolving into a high-tech industry, with AI and big data-driven systems playing a central role. Integrated systems will connect farm machinery and management, transforming farming practices. Educating farmers about technology and optimizing device use are crucial challenges. Younger farmers show more acceptance of technology [65], but safety concerns hinder the widespread adoption of autonomous machines. AI aids in species domestication and requires investment in minor agricultural sectors [66]. Safety measures are vital for IoT devices, particularly in controlled environments like greenhouses [67]. Challenges in sensor development and robotics require multidisciplinary collaboration, aiming for cost-effective and efficient agricultural robots [68, 69].

The combined impact of AI, the IoT, and robotics is expected to enhance quality and productivity in agriculture. Encouraging R&D in this field and government investment in agricultural research is vital for comprehensive farm management integration and ensuring food security. Some of the challenges associated with the integration of AI and their plausible solutions are summarized in Table 1.2.

1.7 FUTURE PROSPECTS

In smart agriculture, the fusion of AI and advanced technologies revolutionizes traditional farming, enhancing efficiency and sustainability. AI facilitates disease detection, automated weed management, and climate-resilient strategies like precision irrigation. It optimizes supply chains, crop genetics, robotic harvesting, and livestock monitoring while promoting energy efficiency. Blockchain ensures food safety and traceability, while collaborative platforms empower farmers. AI-driven drone technology enables precise, real-time data collection and targeted interventions, ushering in a transformative era of precision farming. This synergy promises a more efficient, eco-friendly, and economically viable future for agriculture.

Table 1.2 Summary of Challenges Associated with AI and their Plausible Solutions

S. No.	*Challenge*	*About*	*Plausible Solution*
1.	Increased Food Demand and Population Growth	• A growing global population requires higher quantities of high-quality food. • Rapid population rise necessitates enhanced agricultural production.	• AgTech Innovation • Sustainable Intensification
2.	Yield and Quality Enhancement	• The application of technology is essential to increase both yield and quality. • Agricultural production needs to be boosted through technological advancements.	• Advanced Crop Breeding • Smart Pest and Disease Management
3.	Technology Disparity among Farmers	• Only 2% of farmers with access to modern technology achieve better yield and quality. • Majority struggle due to lack of access to advanced technology.	• Technology Outreach and Training • Digital Access and Infrastructure
4.	Global Discrepancy in Yield	• Developed countries like Australia and Europe achieve significantly higher yields due to technology adoption. • New technology and equipment lead to exponential yield growth in these regions.	• Technology Transfer and Knowledge Sharing • Localized Adaptation of Best Practices
5.	A High-Tech Future of Agriculture	• Agriculture's future is predicted to be technology-driven with AI and big data network systems. • Farm machinery and management will be closely integrated from planting to production.	• Education and Training • Infrastructure and Access
6.	Challenges in Technology Adoption	• Educating farmers about technological devices is a primary challenge. • Building an information base from individual farms for future device optimization is crucial.	• Localized Training and Support • Demonstration Farms and Pilot Projects

Table 1.2 (Cont.)

S. No.	*Challenge*	*About*	*Plausible Solution*
7.	Fear of Labor Displacement	• Concerns about automation and technology replacing traditional labour persists among farmers. • Younger farmers are more open to technology adoption compared to older generations.	• Reskilling and Upskilling • Reskilling and Upskilling
8.	Autonomous Machinery Safety	• Safety concerns hinder the acceptance of autonomous machinery like tractors. • Future developments require improved sensors and control technologies.	• Robust Safety Standards and Regulations • Sensor-Based Safety Systems
9.	IoT Technology and Safety	• IoT devices are widely used in open agricultural environments. • Safety measures are essential for IoT devices exposed to harsh conditions in greenhouses.	• Secure Network Infrastructure • Environmental Protection Measures
10.	Security Challenges in IoT-Based Agriculture	• Hacking and network disruptions are threats to IoT-based agriculture. • Robust security protocols and data loss planning are necessary.	• Multi-Layered Security Protocols • Education and Training
11.	Spatial and Resolution Data Challenges	• Measuring spatial and resolution data is difficult due to extreme variations. • Innovative analytical methods are required to extract knowledge from diverse data layers.	• Advanced Data Fusion Techniques • Remote Sensing and Imaging Technologies
12.	Robotics Speed and Accuracy	• Enhancing robot speed and accuracy for agricultural applications is a key challenge. • Limited funding obstructs progress in this research area.	• Advanced Robotics Design • Real-Time Data Processing and Feedback
13.	Multidisciplinary Approach in Robot Development	• Developing efficient agricultural robots requires a multidisciplinary approach. • Integration of deep learning, intelligent systems, control, engineering, and more is necessary.	• Collaborative Research and Development • Integrated Design Thinking

(continued)

Table 1.2 (Cont.)

S. No.	*Challenge*	*About*	*Plausible Solution*
14.	AI, IoT, and Robotics Integration	• Integrating AI, IoT, and robotics is crucial for solving agricultural challenges. • Comprehensive technology integration is needed for efficient farm management.	• Unified Platform Development • Interdisciplinary Training and Expertise
15.	Research and Development Investment	• Governments should invest in agricultural research and development. • Encouraging research is vital for improving agricultural practices and technology adoption.	• Public-Private Partnerships • Government Funding and Incentives

REFERENCES

1. Elbaşi, E., Mostafa, N., Al-Arnaout, Z., Zreikat, A. I., Cina, E., Varghese, G., Shdefat, A. Y., Topcu, A. E., Abdelbaki, W., Mathew, S., & Zaki, C. (2023). Artificial intelligence technology in the Agricultural sector: A systematic literature review. *IEEE Access*, 11, 171–202.
2. Subeesh, A., & Mehta, C. (2021). Automation and digitization of agriculture using artificial intelligence and internet of things. *Artificial Intelligence in Agriculture*, 5, 278–291. https://doi.org/10.1016/j.aiia.2021.11.004
3. Shaikh, T. A., Rasool, T., & Lone, F. R. (2022). Towards leveraging the role of machine learning and artificial intelligence in precision agriculture and smart farming. *Computers and Electronics in Agriculture*, 198, 107119.
4. World Food Programme. *Global Report on Food Crises* (vol. 2022, p. 277). Rome, Italy: World Food Programme, 2022.
5. Bannerjee G., Sarkar U., Das S., & Ghosh I. (2018). Artificial intelligence in agriculture: A literature survey. *International Journal of Scientific Research in Computer Science Applications and Management Studies*, 7(3), 1–6.
6. Zualkernan, I. A., Abuhani, D. A., Hussain, M., Khan, J., & ElMohandes, M. (2023). Machine learning for precision agriculture using imagery from Unmanned Aerial Vehicles (UAVs): A survey. *Drones*, 7(6), 382.
7. Ayed, R. B., & Hanana, M. (2021). Artificial intelligence to improve the food and agriculture sector. *Journal of Food Quality*, 2021, 1–7.
8. Tsouros, D. C., Bibi, S., & Sarigiannidis, P. (2019). A review on UAV-based applications for precision agriculture. *Information*, 10(11), 349.
9. FAO. *Declaration of the World Summit on Food Security*. Rome, Italy: FAO, 2009.
10. Nagaraja, G. S., Soppimath, A. B., Soumya, T. M., & Abhinith, A. (2019). IoT based smart agriculture management system. *Proceedings of the 4th International Conference on Computer Systems and Information Studies*, EEEXplore, Location: Bengaluru, India

11. FAO. News article: New standards to curb the global spread of plant pests and diseases. www.fao.org/news/story/en/item/1187738/icode/
12. Khakimov, A. A., Salakhutdinov, I. B., Omolikov, A., & Utaganov, S. B. U. (2022). Traditional and current-prospective methods of agricultural plant diseases detection: A review. *IOP Conference Series: Earth and Environmental Science*, 951(1), 012002.
13. Tigges, J., Lakes, T., & Hostert, P. (2013). Urban vegetation classification: Benefits of multitemporal RapidEye satellite data. *Remote Sensing of Environment*, 136, 66–75.
14. Prange, R. K. (2012). Pre-harvest, harvest and post-harvest strategies for organic production of fruits and vegetables. *Acta Horticulturae*, 933, 43–50.
15. Mitchell, J. P., Shennan, C., Grattan, S. R., & May, D. M. (1991). Tomato fruit yields and quality under water deficit and salinity. *Journal of the American Society for Horticultural Science*, 116(2), 215–221.
16. Atay, E., Hucbourg, B., Drevet, A., & Lauri, P. (2017). Investigating effects of over-irrigation and deficit irrigation on yield and fruit quality in Pink Ladytm 'Rosy Glow' Apple. *Acta Scientiarum Polonorum-hortorum Cultus*, 16(4), 45–51.
17. Li, X., Ba, Y., Zhang, M., Nong, M., Yang, C., & Zhang, S. (2022). Sugarcane nitrogen concentration and irrigation level prediction based on UAV multispectral imagery. *Sensors*, 22(7), 2711.
18. Aden, S. T., Bialas, J., Champion, Z., Levin, E., & McCarty, J. L. (2014). Low cost infrared and near infrared sensors for UAVs. *The International Archives of the Photogrammetry, Remote Sensing and Spatial Information Sciences*, XL-1, 1–7.
19. Iowa Ag Literacy. *Why Do They Do That?—Estimating Yields; Iowa Agriculture Literacy*. Des Moines, IA, USA: West, 2019.
20. Altalak, M., Uddin, M. A., Alajmi, A., & Rizg, A. (2022). Smart agriculture applications using deep learning technologies: A survey. *Applied Sciences*, 12(12), 5919.
21. Tuấn, P. A., Sun, M., Nguyen, T., Park, S., & Ayele, B. T. (2019). Molecular mechanisms of seed germination, in Sprouted Grains Nutritional Value, Production and Applications, Eds Hao Feng, Boris Nemzer and Jonathan W. DeVries, Woodhead Publishing and AACC International Press, Cambridge, England. Elsevier (pp. 1–24).
22. Arah, I. K., Amaglo, H., Kumah, E. K., & Ofori, H. (2015). Preharvest and postharvest factors affecting the quality and shelf life of harvested tomatoes: A mini review. *International Journal of Agronomy*, 2015, 1–6.
23. M. E. Barbercheck, Ecological understanding of insects in organic farming systems: How insects damage plants. eOrganic, 1446. https://eorganic.org/node/3151.
24. Sharma, A., Kumar, V., Shahzad, B., Tanveer, M., Sidhu, G. P. S., Handa, N., Kohli, S. K., Yadav, P., Bali, A. S., Parihar, R. D., Dar, O. I., Singh, K. C., Jasrotia, S., Bakshi, P., Ramakrishnan, M., Kumar, S., Bhardwaj, R., & Thukral, A. K. (2019). Worldwide pesticide usage and its impacts on ecosystem. *SN Applied Sciences*, 1(11).

25. Singh, P., Pandey, P. C., Petropoulos, G. P., Pavlides, A., Srivastava, P. K., Koutsias, N., Deng, K. A. K., & Bao, Y. (2020). Hyperspectral remote sensing in precision agriculture: Present status, challenges, and future trends, in Hyperspectral Remote Sensing, Theory and Applications, Eds Prem Chandra Pandey, Prashant K. Srivastava, ... George P. Petropoulos, pp. 121–146, Elsevier.
26. Cisternas, I., Velásquez, I., Caro, Á., & Rodríguez, A. S. (2020). Systematic literature review of implementations of precision agriculture. *Computers and Electronics in Agriculture*, 176, 105626.
27. Hosseiny, B., Rastiveis, H., & Homayouni, S. (2020). An automated framework for plant detection based on deep simulated learning from Drone imagery. *Remote Sensing*, 12(21), 3521.
28. Patel, G. S., Rai, A., Das, N. N., & Singh, R. P. (2023). *Smart Agriculture: Emerging Pedagogies of Deep Learning, Machine Learning and Internet of Things, CRC Press, London.*
29. Gharaei, A., Karimi, M., & Shekarabi, S. a. H. (2019). An integrated multi-product, multi-buyer supply chain under penalty, green, and quality control polices and a vendor managed inventory with consignment stock agreement: The outer approximation with equality relaxation and augmented penalty algorithm. *Applied Mathematical Modelling*, 69, 223–254.
30. Raouhi, E. M., Lachgar, M., Hrimech, H., & Kartit, A. (2023). Unmanned aerial vehicle-based applications in smart Farming: A systematic review. *International Journal of Advanced Computer Science and Applications*, 14(6), pp 1150–1165.
31. Mathushika, M. J., Vinushayini, R., & Gomes, C. (2022). *Smart Farming Using Artificial Intelligence, the Internet of Things, and Robotics: A Comprehensive Review* in Artificial Intelligence and Smart Agriculture Technology, eds Utku Kose, V. B. Prasath, M. Mondal, Prajoy Podder, Subrato Bharati, (pp. 1–19). Auerbach Publications eBooks, New York.
32. Talaviya, T., Shah, D. N., Patel, N., Yagnik, H., & Shah, M. (2020). Implementation of artificial intelligence in agriculture for optimisation of irrigation and application of pesticides and herbicides. *Artificial Intelligence in Agriculture*, 4, 58–73.
33. Jha, K., Doshi, A., Patel, P., & Shah, M. (2019). A comprehensive review on automation in agriculture using artificial intelligence. *Artificial Intelligence in Agriculture*, 2, 1–12.
34. TeKrony, D. M., & Egli, D. B. (1991). Relationship of seed vigor to crop yield: A review. *Crop Science*, 31(3), 816–822.
35. Colmer, J., O'Neill, C. M., Wells, R., Bostrom, A., Reynolds, D. B., Websdale, D., Shiralagi, G., Wei, L., Lou, Q., Cornu, T. L., Ball, J., Renema, J., Andaluz, G. F., Benjamins, R., Penfield, S., & Zhou, J. (2020). SeedGerm: A cost-effective phenotyping platform for automated seed imaging and machine-learning based phenotypic analysis of crop seed germination. *New Phytologist*, 228(2), 778–793.
36. Kumar, R., Singh, M. P., Kumar, P., & Singh J. P. (2015). Crop selection method to maximize crop yield rate using machine learning technique.

2015 International Conference on Smart Technologies and Management for Computing (pp. 138–145). Avadi, India: Communication, Controls, Energy and Materials (ICSTM).

37. Lakhiar, I. A., Gao, J., Syed, T. N., Chandio, F. A., Buttar, N. A., & Qureshi, W. (2018). Monitoring and control systems in agriculture using intelligent sensor techniques: A review of the aeroponic system. *Journal of Sensors*, 2018, 1–18.
38. Facchinetti, D., Santoro, S., Galli, L. E., Fontana, G., Fedeli, L., Parisi, S., Bonacchi, L. B., Šušnjar, S., Salvai, F., Coppola, G., Matteucci, M., & Pessina, D. (2021). Reduction of pesticide use in fresh-cut Salad production through Artificial Intelligence. *Applied Sciences*, 11(5), 1992.
39. Van Klompenburg, T., Kassahun, A., & Çatal, Ç. (2020). Crop yield prediction using machine learning: A systematic literature review. *Computers and Electronics in Agriculture*, 177, 105709.
40. Partel, V., Kakarla, S. C., & Ampatzidis, Y. (2019). Development and evaluation of a low-cost and smart technology for precision weed management utilizing artificial intelligence. *Computers and Electronics in Agriculture*, 157, 339–350.
41. Mulla, D. J. (2013). Twenty five years of remote sensing in precision agriculture: Key advances and remaining knowledge gaps. *Biosystems Engineering*, 114(4), 358–371.
42. Zhang, C., & Kovacs, J. M. (2012). The application of small unmanned aerial systems for precision agriculture: A review. *Precision Agriculture*, 13(6), 693–712.
43. Gupta, S. G., Ghonge, M. M., & Jawandhiya, P. M. (2013). *Review of Unmanned Aircraft System (UAS)*. Social Science Research Network. 2(4), India.
44. Oettershagen, P., Stastny, T., Hinzmann, T., Rudin, K., Mantel, T., Melzer, A., Wawrzacz, B., Hitz, G., & Siegwart, R. (2017). Robotic technologies for solar-powered UAVs: Fully autonomous updraft-aware aerial sensing for multiday search-and-rescue missions. *Journal of Field Robotics*, 35(4), 612–640. https://doi.org/10.1002/rob.21765
45. Zhao, T., Niu, H., Rosa, E., Doll, D., Wang, D., & Chen Y.Q. (2018). Tree canopy differentiation using instance-aware semantic segmentation. *Proceeding of 2018 ASABE Annual International Meeting, Detroit, MI, USA, Jul. 29* (vol. 2018, p. 1). St. Joseph, MI, USA: American Society of Agricultural and Biological Engineers, Aug. 1.
46. Torres-Sánchez, J., De Castro, A. I., Peña, J. M., Jiménez-Brenes, F. M., Arquero, O., Lovera, M., & López-Granados, F. (2018). Mapping the 3D structure of almond trees using UAV acquired photogrammetric point clouds and object-based image analysis. *Biosystems Engineering*, 176, 172–184.
47. Sa, I., Popović, M., Khanna, R., Chen, Z., Lottes, P., Liebisch, F., Nieto, J., Stachniss, C., Walter, A., & Siegwart, R. (2018). WeedMap: A Large-Scale semantic weed mapping framework using aerial multispectral imaging and deep neural network for precision farming. *Remote Sensing*, 10(9), 1423.

48. Bonotis, P., Tsouros, D. C., Smyrlis, P. N., Tzallas, A. T., Γιαννακέας, N., Glavas, E., & Tsipouras, M. G. (2019). Automated assessment of pain intensity based on EEG signal analysis. *2019 IEEE 19th International Conference on Bioinformatics and Bioengineering (BIBE),* IEEExplore.
49. Tarca, A. L., Carey, V. J., Chen, X., Romero, R., & Drăghici, S. (2007). Machine learning and its applications to biology. *PLOS Computational Biology*, 3(6), e116.
50. Leica Geosystems. *ERDAS Imagine*. Atalanta, GA, USA: Leica Geosystems, 2004.
51. Gnädinger, F., & Schmidhalter, U. (2017). Digital counts of maize plants by unmanned aerial vehicles (UAVs). *Remote Sensing*, 9(6), 544.
52. Fan, X., Kawamura, K., Xuan, T. D., Yuba, N., Lim, J., Yoshitoshi, R., Minh, T. N., Kurokawa, Y., & Obitsu, T. (2017). Low-cost visible and near-infrared camera on an unmanned aerial vehicle for assessing the herbage biomass and leaf area index in an Italian ryegrass field. *Grassland Science*, 64(2), 145–150.
53. Marino, S., & Alvino, A. (2018). Detection of homogeneous wheat areas using multi-temporal UAS images and ground truth data analyzed by cluster analysis. *European Journal of Remote Sensing*, 51(1), 266–275.
54. Baatz. *eCognition Professional User Guide 4*. Munich, Germany: ADefiniens Imaging, 2004.
55. Tetracam, Inc. *ADC Users Guide, V.2.3*. Chatsworth, CA, USA: Tetracam Inc., 2011.
56. Zou, K., Chen, X., Zhang, F., Zhou, H., & Zhang, C. (2021). A field weed density evaluation method based on UAV imaging and modified U-Net. *Remote Sensing*, 13(2), 310.
57. Shahi, T. B., Xu, C., Neupane, A., Fresser, D., O'Connor, D., Wright, G. C., & Guo, W. (2023). A cooperative scheme for late leaf spot estimation in peanut using UAV multispectral images. *PLOS ONE*, 18(3), e0282486.
58. Xue, X., Lan, Y., Sun, Z., Chang, C., & Hoffmann, W. C. (2016). Develop an unmanned aerial vehicle based automatic aerial spraying system. *Computers and Electronics in Agriculture*, 128, 58–66.
59. Quebrajo, L., Pérez-Ruiz, M., Pérez-Urrestarazu, L., Martínez, G., & Egea, G. (2018). Linking thermal imaging and soil remote sensing to enhance irrigation management of sugar beet. *Biosystems Engineering*, 165, 77–87.
60. Jung, J., Maeda, M. M., Chang, A., Landivar, J., Yeom, J., & McGinty, J. A. (2018). Unmanned aerial system assisted framework for the selection of high yielding cotton genotypes. *Computers and Electronics in Agriculture*, 152, 74–81.
61. Kerkech, M., Hafiane, A., & Canals, R. (2018). Deep leaning approach with colorimetric spaces and vegetation indices for vine diseases detection in UAV images. *Computers and Electronics in Agriculture*, 155, 237–243.
62. Bah, M. D., Hafiane, A., & Canals, R. (2017). Weeds detection in UAV imagery using SLIC and the hough transform. *Conference: 2017 Seventh International Conference on Image Processing Theory, Tools and Applications (IPTA)*, IEEEXplore, Montreal, QC, Canada
63. Eurostat. *Agriculture, Forestry and Fishery Statistics*. Luxembourg: Eurostat, 2018.

64. European Commission. (2019). "Commission delegated" Regulation (EU) 2019/945 of 12 March 2019 on unmanned aircraft systems and on third-country operators of unmanned aircraft systems. *Official Journal of European Union*, L152, 1–40.
65. Ayaz, M., Uddin, M. A., Sharif, Z., Mansour, A., & Aggoune, E. M. (2019). Internet-of-Things (IoT)-based smart agriculture: Toward making the fields talk. *IEEE Access*, 7, 129551–129583.
66. Zha. (2020). Artificial intelligence in agriculture. *Journal of Physical Conference Sciences*, 1693(1), 012058.
67. Harris, D. R., & Fuller, D. Q. (2014). Agriculture: Definition and overview. Springer (pp. 104–113). Smith, C. (ed) Encyclopaedia of Global Archaeology. New York.
68. Farooq, M. S., Riaz, S., Abid, A., Umer, T., & Zikria, Y. B. (2020). Role of IoT technology in agriculture: A systematic literature review. *Electronics*, 9(2), 319.
69. Paul, P., , K, R. Saavedra., Aithal, P. S., Sinha, R. R., & Aremu, B. (2020). *Agro Informatics Vis-à-Vis Internet of Things (IoT) Integration & Potentialities—An Analysis*. Social Science Research Network, 7(1), 13-20, New Delhi, India.

Chapter 2

Enhancing Food Quality Assurance

Leveraging the IoT for Smart Agriculture Monitoring from Pre to Post-Harvest

C.V. Suresh Babu and A. Sanjai Das

2.1 INTRODUCTION

This chapter aims to explore the implementation of the Internet of Things (IoT) in monitoring food quality during both pre- and post-harvest stages in the context of smart agriculture. The need to ensure food safety, minimize wastage, and maintain the highest possible quality of food products throughout the entire supply chain presents significant challenges to the agriculture and food sectors. Leveraging IoT technologies for real-time data collection, analysis, and control can offer innovative solutions to address these challenges and revolutionize the way we produce, store, and transport food.

2.1.1 Overview of IoT in Smart Agriculture

The adoption of the Internet of Things (IoT) in smart agriculture has emerged as a game-changing answer to the problems posed by observing food quality both before and after harvest. The agriculture and food sectors may transform their operations by utilizing IoT technology, assuring food safety, cutting waste, and maintaining enhanced capacity along the whole distribution chain [1].

Sensors enabled by the Internet of Things may collect crucial environmental data on crop development and health, increasing agricultural operations and allowing farmers to make educated decisions. These sensors can also identify early warning signs of disease and pests, decreasing the need for pesticides and maintaining healthy crop development.

Post-harvest monitoring is critical for preserving food quality since IoT sensors keep items safe and fresh during storage and transportation. When deviations from ideal criteria occur, real-time notifications are generated, reducing spoilage and waste. IoT technologies can make cold chain management easier, ensuring perishable commodities maintain quality and shelf life. Through the IoT, device data, data analytics and decision support are possible, allowing stakeholders to make informed decisions and enhance food

 DOI: 10.1201/9781003676102-2

quality and safety. End-to-end food product traceability across the supply chain is enabled by the integration of the IoT and blockchain technology, giving precise information on the origin, manufacturing, and management of food.

Smart agriculture powered by the IoT enhances efficiency and sustainability by reducing waste of water, fertilizer, and pesticides, resulting in higher crop yields and resource efficiency, making the agricultural business more environmentally friendly and sustainable.

The use of linked devices, sensors, and data analytics technology to enhance different agricultural operations is known as the Internet of Things (IoT) in agriculture. Farmers may gather real-time data from their fields, animals, and other agricultural assets by incorporating the IoT into their farming methods. This enables them to make data-driven choices and improve overall efficiency, production, and sustainability in the agricultural sector.

The IoT in agriculture refers to internet-connected gadgets, sensors, actuators, and smart devices that collect and transmit data. These gadgets gather data on environmental factors including temperature, humidity, soil moisture, light levels, and weather patterns as well as equipment efficiency and animal health indicators. The device layer, connectivity layer, and application layer are the three main parts of the implementation of IoT technologies, from data collection in the 'Things,' through application management in designated IoT development platforms. This attempts to offer shipping travels, supply chain operations, environmental monitoring, and food identification [2]. The collected data is then transmitted to cloud-based platforms or regional systems for storage, processing, and analysis. The technology processes the data, drawing insightful conclusions and spotting trends for decision-making. The data supports farmers in addressing crop health, resource allocation, pest control, and irrigation timing.

IoT-enabled devices can automate agricultural activities, such as modifying irrigation systems and improving greenhouse climate management. Precision farming allows farmers to customize their operations to specific crop or livestock demands, minimizing waste and maximizing resource efficiency. The IoT can also be integrated into the agricultural supply chain to monitor produce transportation, storage, and distribution, ensuring food quality and safety from farm to table.

2.1.2 IoT Applications in Agriculture and Food Industry

In recent years, IoT applications have had a significant positive impact on agriculture and the food business. By incorporating smart devices, sensors, and data analytics, the IoT has modernized outdated systems and increased productivity, efficiency, and sustainability. One common application for IoT sensors in agriculture is precision farming, which monitors soil moisture, temperature, and nutrient levels. Not only can soil provide nourishment, but in addition, it serves as the planet's natural water filter and is the second-largest carbon storage system after the oceans. The significance of soil is equally perplexing. For agriculture to succeed, rich soil is necessary. For the plants to flourish, the soil must have nutrients. The soil provides nutrients for our meals. When the soil is suitably rich, it will also help in boosting yield rates [3].

Real-time data is sent to farmers to optimize fertilization and irrigation, resulting in higher crop yields and less resource waste. Another crucial IoT use is livestock monitoring, which enables farmers to keep tabs on the behavior and health of their livestock, enabling early disease identification and enhancing animal welfare. Temperature control is used in IoT-enabled smart greenhouses to facilitate year-round agricultural productivity and energy efficiency.

IoT-powered automated irrigation systems reduce water use by altering water distribution in response to weather and soil conditions.

2.1.3 Role of AI and IoT Integration in Smart Agriculture

Modernizing and improving farming methods depends heavily on the integration of Artificial Intelligence (AI) and the Internet of Things (IoT) in smart agriculture. With the help of automated decision-making and data-driven insights, this potent combination gives farmers greater productivity, sustainability, and profitability [4, 10].

AI algorithms examine the information gathered by IoT sensors placed on cattle, in greenhouses, and in fields. These sensors keep an eye on several variables, including temperature, humidity, crop health, and animal behavior. Precision farming, which allows farmers to adjust irrigation, fertilization, and pest control measures to meet the unique demands of their crops or livestock, is made possible by AI-driven analysis. With this optimization, resource waste is reduced, fewer chemicals are used, and overall yield quality is raised [1].

Using AI and IoT data, predictive analytics may help farmers foresee possible problems like insect outbreaks, disease outbreaks, or unfavourable weather conditions. Farmers may proactively plan their operations and take preventative steps by spotting patterns and trends, minimizing risks and improving output outcomes.

Automated monitoring and control systems are also made possible by the combination of AI and the IoT. Based on real-time data, AI algorithms may remotely control watering and feeding systems and automatically alter parameters in smart greenhouses. This lessens the need for manual intervention, boosts operational effectiveness, and guarantees ideal growth conditions.

2.2 UNDERSTANDING THE FOOD QUALITY CHALLENGES IN THE SUPPLY CHAIN

Food products are the most delicate goods on the market, for they are exposed to risks like inadequate storage, wrong temperature levels, poor air quality, humidity, light and other factors influencing their quality and safety. The examples corroborating this are: with no optimal temperature, bacteria grow uncontrollably, and the light directly influences the quality of a fresh product likewise, without proper storage, goods become perishable. Hence, control equipment that can instantly indicate the temperature is required in every storage space. Participants in the food supply chain such as farmers, suppliers, wholesalers, retailers, and transporters are mandated to provide the right conditions for food items.

The flow of information among the participants is also important, as well as product management, in order to maintain the food quality in the supply chain. The food product supply chain relates to all the processes (production, manufacturing, distribution, sales, and consumption) that describe the flow of foods from the farm (field) to the consumer's table. Each step of the supply chain needs the use of either natural or human resources, and if any link in this chain is put at risk, this has an impact on the entire chain.

To avoid the chain falling apart, namely, the products perishing, it is crucial that participants value certain characteristics of a product, including its origin, legal requirements, and the temperature sensibility set by the producer's declaration. The distributors are required to always assure product traceability using specialized information technology for temperature monitoring, and so forth, and are key players in this. Given that the food sold must be of a specific calibre, a high level of security is required because the supply chain itself is situated in a challenging, dynamic, and time-sensitive environment.

2.2.1 Pre-harvest Factors Affecting Food Quality

The quality of food items is greatly impacted by pre-harvest variables before they are harvested. These elements have a significant role in determining the ultimate nutritional value, flavor, appearance, and market value of agricultural products.

Some of the most important pre-harvest factors influencing food quality are as follows:

Environmental Factors: Weather and climate play a critical role in the development and growth of crops. Extremes in temperature, humidity, and precipitation can all have an impact on the rate of photosynthesis, the absorption of nutrients, and the general health of the plants.

The quality and make-up of the soil have an impact on the amount of nutrients in the crop. The development of healthier plants and improved nutrient quality in the food produced are encouraged by the presence of essential minerals and microbes as well as adequate soil fertility.

Water quality: The source and quality of irrigation water can have an impact on crop health and the accumulation of pollutants in food. Water that has been contaminated with contaminants or microorganisms can pose a risk to food safety.

Management of Pests and Diseases: To avoid crop damage, efficient pest and disease management techniques are essential. Infections and infestations can lower yields and degrade the quality of food.

The genetics of a crop and its variety can affect a crop's taste, texture, nutritional value, and susceptibility to pests and diseases, among other factors. The application of pesticides, herbicides, and fertilizers can all have an impact on the quality of food. Crops that have been harvested may contain chemical residues as a result of excessive or improper use of agrochemicals.

2.2.1.1 *Environmental Conditions and their Impact on Crop Health*

Crop health, growth, and total agricultural productivity are significantly impacted by environmental factors. Many factors affect the health and quality of agricultural products, and there is a complicated and ever-changing relationship between crops and their surroundings. Understanding these environmental factors is essential for putting sustainable agricultural strategies into practice and preserving the security of the world's food supply.

Following are some important environmental elements and how they affect crop health:

2.2.1.1.1 Temperature

- Plant metabolism may be accelerated by high temperatures, which can result in quicker growth and more water. Extreme heat stress, however, can harm plant tissues, slow down photosynthesis, and have an impact on fruit setting and flowering.
- Low temperatures, especially for sensitive crops, can impede crop development, postpone maturity, and increase the risk of frost damage.

2.2.1.1.2 Rainfall and the Availability of Water

- The growth of crops depends on their being enough rainfall to give plants the water and nutrients they need. A lack of rain can cause drought stress, which can restrict growth and lower harvests.

- Increased disease susceptibility, nutrient leaching, and root suffocation can all be effects of excessive rainfall or waterlogging.

2.2.1.1.3 Humidity

- Excessive humidity can harm plants' health and decrease output by fostering the growth of fungi and diseases like powdery mildew.
- Increased transpiration rates brought on by low humidity may cause crops to become water stressed and wilt.

2.2.1.1.4 Light

- Photosynthesis, which enables plants to transform carbon dioxide and water into sugars and oxygen, depends on sunlight. High-quality product and healthy development are encouraged by enough light exposure.
- Poor crop quality, extended stems, and impaired fruit growth are all consequences of insufficient light.

2.2.1.1.5 Soil Type

- Plant health is directly impacted by nutrient content and soil composition. Nutrient shortages brought on by poor soil fertility can have an impact on crop development and production.
- Root growth can be hampered and the danger of waterlogging is increased by compacted soil and inadequate drainage.

2.2.1.1.6 High-quality Air

- Ozone and sulfur dioxide are two air pollutants that can harm plant tissues, slow down photosynthesis, and lower agricultural yields.
- Pollen contamination brought on by air pollution can impact the viability of the reproductive processes of some crops and of seed quality.

2.2.1.1.7 Disease and Pest Pressure

- Pest and disease activity and prevalence are influenced by environmental factors. For instance, warm and humid weather can hasten the development and spread of diseases and insect pests.

2.2.1.2 Pest and Disease Monitoring for Early Intervention

Integrative Pest Management (IPM) in agriculture is a crucial aspect that involves the early detection of and intervention against pests and diseases.

This helps prevent outbreaks, reduces crop loss, and reduces the use of chemical pesticides.

Effective monitoring ensures sustainable crop output and allows farmers to use focused, ecologically friendly management methods. Techniques for pest and disease monitoring include visual inspections, traps and lures, sticky traps, automated sensors and IoT devices, remote sensing, and disease forecasting models. Visual inspections can detect insect infestations or disease symptoms, while traps and lures help monitor insect numbers and identify potential pest outbreaks. Sticky traps help determine possible pest outbreaks and assess insect numbers.

In order to provide early warning indications for intervention, automated sensors and IoT devices may gather real-time data on ambient conditions, insect presence, and sickness frequency. By considering decentralized data management, consensus methods, and mining activities, the use of Blockchain technology may monitor and trace food in a farm-to-table manner with improved transparency across the supply chain. Science has examined the viability, implementation, and difficulties of integrating Blockchain with the IoT [5, 6, 7, 8]. Remote sensing methods, such as satellite photography and drones equipped with multispectral sensors, can detect changes in crop health and stress caused by pests or diseases. Disease forecasting models can also anticipate disease outbreaks, allowing farmers to implement preventive measures before they spread.

Early intervention techniques, such as the following, can be used once pests or illnesses have been identified through monitoring:

- Cultural practices: Planting resistant cultivars and rotating crops can assist in reducing the stresses brought on by pests and diseases.
- Beneficial insects: Creatures can be introduced to naturally manage pest populations through biological control.
- Mechanical Control: Localized infestations can be controlled using physical techniques like handpicking bugs or removing harmed plant sections.
- Targeted Pesticide Use: If necessary, only the afflicted regions can be given applications of low-impact, selective pesticides, reducing the effects on the ecosystem.

Farmers may accomplish successful pest control while protecting ecosystem health, cutting costs, and assuring sustainable agricultural practices by integrating pest and disease monitoring with early intervention measures.

2.2.2 Post-harvest Challenges in Maintaining Food Quality

In order to avoid food spoilage, cut down on food waste, and guarantee that customers are provided with safe and nourishing food, it is important to address the problems associated with maintaining food quality throughout the post-harvest stage.

Following harvest, maintaining food quality presents a number of significant issues.

Temperature Control: Maintaining the right temperature is essential for keeping food goods fresh and limiting the growth of bacteria that cause spoilage. Accelerated degradation and microbiological development might result from inadequate cooling or refrigeration during storage and transit.

Controlling Moisture: This is important because too much moisture can encourage the growth of mildew and deterioration, while not enough moisture can make fresh food wilt and lose its texture. To maintain the quality of food, the humidity must be balanced properly.

Post-Harvest Illnesses: Certain infections can induce post-harvest illnesses, which can diminish the nutritive content, flavour, and appearance of fruits and vegetables. It is essential to follow good hygiene and sanitation procedures to stop the spread of infections.

Techniques for Harvesting: The way food is harvested can affect its quality. The crops may sustain physical, bruising, or mechanical harm as a result of improper techniques.

2.2.2.1 Temperature and Humidity Control During Storage and Transportation

Controlling temperature and humidity during storage and transportation are essential components of post-harvest food management. Keeping things in top shape helps to extend shelf life, maintain quality, and guarantee food safety. A reduced danger of deterioration and mold growth is achieved by maintaining proper humidity and cooling levels, which prevent the formation of spoilage bacteria and enzymes.

Controlled temperatures in cold storage facilities slow down metabolic activity in fresh produce, delaying ripening and increasing product life. Also, ideal humidity levels reduce moisture loss and fruit and vegetable withering.

To prevent food waste, improve the efficiency of the supply chain, and provide customers with high-quality, safe, and nutrient-rich food, it is crucial to put appropriate temperature and humidity control mechanisms into place.

For handling and transportation, the quality of the produce is preserved when it is handled carefully during harvesting and transportation to avoid physical harm, bruising, and contamination.

Post-harvest storage conditions are essential for maintaining food quality. Food quality after harvesting depends on setting up suitable storage conditions. The shelf life and freshness of the produce can be affected by variables like temperature, humidity, and ventilation.

Rigorous testing and grading techniques are frequently used to ensure consistent food quality. Yet, variation in product quality may result from the absence of defined quality evaluation procedures.

For perishable food items, maintaining an uninterrupted cold chain from the farm to the customer is crucial. Any interruption in the cold chain can cause food to deteriorate and spoil quickly.

2.2.2.2 *Food Spoilage and Wastage Issues*

The global food system faces considerable obstacles from food deterioration and waste. Spoilage is a microbiological, enzymatic, or physical process that impairs the quality and safety of food. Post-harvest losses are a result of poor handling, handling techniques, and transportation procedures. Food waste can also happen throughout several phases of manufacturing, processing, distribution, and consumption.

Waste can be caused by overproduction, inadequate storage, aesthetic defects, and consumer behavior. Food waste has a negative impact on the environment as well as wasting precious resources and escalating food poverty. The Internet of Things (IoT) is a new paradigm that integrates the physical and digital worlds to address pressing issues in daily living. The IoT foresees the connectivity of numerous intelligent items and devices in our environment that have the capacity to gather processes and transmit data [9]. Better post-harvest management, effective supply chains, consumer education, and sustainable practices are required to address these concerns to decrease food loss and increase food security.

Pathogens and Pests: Post-harvest storage spaces can serve as breeding grounds for diseases and pests. Ineffective pest management practices can cause infestations, which can cause food losses and quality degradation.

Ripening and Maturity: Crops have the best flavour, texture, and nutritional value when they are harvested at the right point in their maturation. Food quality may suffer from harvesting occurring too early or too late.

Senescence and Ripening: After harvest, the ripening and senescence rates of various fruits and vegetables differ. To avoid over-ripening or early degradation, it is essential to control the dates of harvest and storage conditions.

2.3 IOT SOLUTIONS FOR PRE-HARVEST FOOD QUALITY MONITORING

Due to its potential to increase food safety, decrease waste, and improve all aspects of food preservation techniques, the Internet of Things (IoT) has attracted a lot of attention in the field of pre-harvest food quality monitoring. When used in the food business, the IoT provides real-time monitoring of environmental factors that impact the quality and freshness of fruits and vegetables during storage. The IoT aims to link various items and things through the internet.

Researchers and specialists are designing and implementing IoT-based solutions for monitoring the ambient conditions in which food products are stored as a result of the fast growth of IoT technology. One such recommended system makes use of a number of sensors, such as pH sensors to measure the pH levels of the food, temperature sensors to keep track of the temperature of the food and its salt content, and gas sensors to look for the release of certain gases, such as ethanol, which signify food decomposition.

Pre-harvest food quality monitoring is important because it can spot food degradation before it manifests itself in obvious ways. The suggested IoT system uses sensors attached to a microcontroller to detect and monitor the particular gases that food deterioration generates (such as Arduino). The device can assess the freshness of food products and give real-time feedback by gathering information on temperature, moisture, and gas levels.

Such food quality monitoring systems are even more necessary in nations like India, where contaminated foodborne diseases frequently result in hospitalizations and fatalities. IoT solutions can guarantee that food safety requirements are upheld across the supply chain, from the farm to the consumer, by continuously monitoring.

By using IoT technology in the food industry, intelligent food storage facilities may be created. These storage facilities feature sensor nodes that compile data from a variety of environmental situations. The examination of the aggregated data, followed by the generation of a consolidated output, helps in making intelligent judgments on the quality and freshness of the stored food goods.

The suggested IoT system may also be improved by incorporating an Android mobile application, which would facilitate user engagement and provide simple access to real-time data on food freshness. These tools can help consumers make better food decisions and lower their risk of contracting foodborne diseases.

2.3.1 Sensing and Monitoring Technologies for Agricultural Applications

The agriculture industry is being significantly impacted by the Internet of Things (IoT) and Artificial Intelligence (AI); two of the most disruptive technologies of our time. IoT sensors may be used to gather information on a number of variables, including weather patterns, crop health, and soil conditions. Then, by using this information, agricultural techniques like irrigation, fertilization, and pest management may be improved.

IoT sensor data may be analyzed by AI to find patterns in the data that can be utilized to enhance agricultural methods. AI may be used, for instance, to forecast crop yields, spot pests, and illnesses early on, and improve irrigation techniques. Due to the confluence of the IoT and AI, farmers now have new alternatives to improve their operations and boost yields.

Here are some examples of how the IoT and AI are used in agriculture:

- Precision Farming: IoT sensors are used in precision farming to gather information on the health of the crops, the soil, and the weather. The farm is then precisely mapped out using this data so that irrigation, fertilization, and pest control can all be improved.
- Agricultural Monitoring: IoT sensors may be used to track the health, stress, and growth of crops. This information may be utilized to enhance crop harvesting and early pest and disease detection.
- Artificial Intelligence (AI) may be used to evaluate weather sensor data and anticipate weather trends. Using this knowledge will improve watering procedures and safeguard crops from severe weather.
- Robotics: Tasks like planting, watering, and harvesting crops may be automated with robots. This can be advantageous in increasing productivity and lowering labor expenses.

Although the application of the IoT and AI in agriculture is still in its infancy, it has the potential to completely transform the way that it is carried out. These innovations can assist farmers in increasing agricultural yields, cutting expenses, and preserving the environment.

Here are a few advantages of using the IoT and AI in agriculture:

- Crop Yields: These will increase as a result of improved irrigation, fertilization, and pest management decisions made by farmers as a result of data collection on soil characteristics, crop health, and weather patterns. Crop yields may rise as a result of this.
- Cost Savings: Farmers may cut back on the quantity of water, fertilizer, and pesticides they use by enhancing agricultural methods. Significant cost reductions may result from this.
- Enhanced Environmental Sustainability: Farmers may lessen their environmental effect by utilizing the IoT and AI. They may, for instance, utilize water more wisely and refrain from using dangerous pesticides.

The expense of these technologies and the lack of data standards are two issues that continue to hinder the application of the IoT and AI in agriculture. But these issues are being resolved, and the IoT and AI in agriculture have a very bright future.

The following are some of the challenges associated with using the IoT and AI in agriculture:

- Cost: AI and Internet of Things (IoT) technology can be pricey, especially for small farms.
- Data Standardization: The way IoT and AI sensors gather and distribute data is not standardized. The integration of data from many sources may become challenging as a result.
- IoT and AI sensors are vulnerable to hacks since they are linked to the internet.

2.3.1.1 Weather Monitoring and Climate Sensors

Modern meteorological and environmental monitoring systems must include sensors for weather and climate. These sensors are made to monitor different weather factors and climatic conditions, giving vital information for climate research, weather forecasts, and decision-making across a range of industries.

Temperature sensors are crucial for weather tracking, measuring air temperature in real-time and are being used in weather stations, cellphones, and IoT devices.

Hygrometers, or humidity sensors, are crucial tools for assessing air moisture levels, aiding in agriculture, interior climate management, and weather forecasting.

Barometric pressure sensors are essential in weather stations and other meteorological apparatus for monitoring air pressure, a crucial indicator of weather changes.

Wind sensors are crucial for navigation, wind energy production, and weather forecasting by measuring wind speed and direction.

By calculating the quantity of rain, snow, or hail that has fallen, precipitation sensors are essential for managing water resources, forecasting floods, and hydrology.

To analyze climatic changes and evaluate the potential for solar energy production, solar radiation sensors monitor solar energy reaching the planet's surface. Technology elements including sensors, mobility, real-time monitoring, connection, and the simplicity of application deployment are all necessary for precision agriculture. The agriculture industry is already participating in the next industrial revolution, and farmers are implementing cutting-edge technologies to fulfil the need for food [10].

UV sensors measure the sun's UV radiation strength, aiding in health risk assessments and monitoring UV levels for environmental investigations.

Evapotranspiration sensors are used in agriculture and water management to optimize irrigation practices by tracking the simultaneous evaporation of soil moisture and plant transpiration [11].

Soil moisture sensors aid in agricultural decision-making, drought monitoring, and irrigation scheduling.

Leaf wetness sensors measure leaf moisture, helping farmers prevent illnesses caused by excessive plant wetness.

Snow depth sensors are essential for managing water resources, predicting avalanches, and monitoring snowpack depth to ensure accurate snow cover measurements.

The Cubic NDIR CO_2 Sensor SRH-40 is an atmospheric gas sensor that measures CO_2 concentrations, providing valuable data for climate research and air quality monitoring.

Cloud height sensors are essential tools for aviation, weather forecasting, and climatology studies because they measure the height of clouds.

Weather radars use radio waves to detect precipitation, track its path and intensity, and alert the public about storms and severe weather conditions.

2.3.1.2 Remote Sensing for Crop Health Assessment

In evaluating crop health, remote sensing is essential because it offers useful data on the health and productivity of crops across vast agricultural areas. It involves gathering information about the electromagnetic radiation reflected or released into the atmosphere by crops incorporating sensors and imaging technology mounted on spacecraft, drones, or aircraft. The different crop-specific areas are then evaluated to use this data.

Crop state estimation necessitates an input of information, such as environmental conditions like air temperature, relative humidity, and rainfall, as well as surface conditions like soil moisture and soil temperature. Remote sensing indices are used, for example, the Soil Adjusted Vegetation Index (SAVI), the Temperature-Vegetation Dryness Index (TVDI), the Normalized Difference Vegetation Index (NDVI), the Water Deficit Index (WDI), the Land Surface Water Index (LSWI), and so forth.

The ability to obtain a global synoptic view of the organization of world is greatly facilitated by remote sensing [12]. Since the advent of satellite data, a variety of classes of selective information on the available resources, including soil humidity, soil use and cover, crop natures and states, and soil type data, can be mined [13]. Some of these are:

Chlorophyll Content: A vital pigment in plants, chlorophyll is responsible for photosynthesis. The chlorophyll content of crops can be estimated through remote sensing, giving information about their physiological state and general health.

Water Stress: By monitoring changes in leaf water content and canopy temperature, remote sensing data can be used to identify water stress in crops. The improvement of irrigation techniques and water management requires this knowledge [11].

Crop Stress Monitoring: Remote sensing is used to keep track of the many stressors that affect crops, including salinity, drought, and heat stress. The application of stress management techniques is made possible by prompt detection.

2.3.2 Data Analytics and Decision Support Systems

Improved IoT-enabled Food Quality Assurance Data analytics and Decision Support Systems (DSS) are integrated into the monitoring of smart agriculture at many phases of the agricultural process, from pre-harvest to post-harvest.

2.3.2.1 *Pre-Harvest Data Analytics*

- Data Gathering: This comes from the IoT sensors and tools that are utilized to gather information on soil quality, weather, crop development, and insect infestations prior to harvest [14].
- Data Processing: The raw data will be subjected to data analytics processes such as data cleansing, integration, and transformation, in order to provide actionable insights. This allows the use of predictive models to anticipate agricultural yields, spot possible disease outbreaks, and optimize planting schedules in the context of predictive analytics.

2.3.2.2 *Post-Harvest Data Analytics*

- These describe how IoT sensors will continue to keep an eye on conditions throughout storage and transit, such as temperature, humidity, and handling. Data analytics can be used to evaluate the quality of harvested crops, find spoilage, and trace the provenance of products.

2.3.2.3 *Decision Support Systems for Pre-Harvest*

- These make decisions regarding irrigation, fertilizer, and pest management based on current data and forecasting models.
- These will send out notifications with warnings and suggestions and offer practical advice to farmers and other agricultural employees.

2.3.2.4 *Decision Support Systems for Post-Harvest*

- These list the ways in which DSS will help with inventory management, storage optimization, and the reduction of food waste.
- DSS will support quality control procedures and make sure that only produce of the highest caliber reaches customers.

2.3.2.5 *Predictive Models for Optimizing Growing Conditions*

In agriculture, predictive models are critical for increasing crop yields and quality. They forecast and optimize growth conditions using data from IoT devices, historical meteorological data, and soil information, allowing farmers to make more educated decisions [14].

Predictive models are used by farmers to plan planting and harvesting dates, monitor soil health and fertility, and manage pests and diseases. Farmers may organize their efforts to prevent bad weather by using historical weather data and climate forecasts. Farmers may make educated judgments about soil amendments and pH modifications if they understand the soil conditions. Predictive models can also predict pest infestations and disease outbreaks, allowing farmers to adopt preventative or targeted treatments.

Crop yields are projected using predictive models based on historical data, allowing farmers to adjust planting density and other factors. They also aid in fertilizer management by measuring crop demands and soil nutrient levels to ensure crops receive the nutrients they require for optimal development. Decision Support Systems (DSS) are agricultural systems that give farmers practical guidance on subjects such as planting, harvesting, and irrigation. IoT sensors and remote sensing technologies improve predictive models and forecast accuracy by providing real-time data on temperature, humidity, and soil moisture.

2.3.2.6 *Early Warning Systems for Pest and Disease Detection*

Agriculture is essential for human survival and the global economy, but pests and diseases pose serious dangers to crop output and food security. Early warning systems that leverage technology and data analysis are critical instruments for identifying, monitoring, and managing epidemics before they cause widespread damage. This paper investigates the significance, components, and advantages of agricultural early warning systems are as follows:

- The Importance of Early Warning Systems: Pest and disease infestations can cause considerable agricultural damage, economic losses, and food shortages. Farmers must notice pests early to take prompt, focused action. Farmers frequently rely on labor-intensive visual inspections in the absence of early warning systems, which may not uncover concerns until they have become entrenched.
- Components of Early Warning Systems: Early warning systems collect data on environmental conditions, insect populations, and disease indicators from a network of data sources that include weather stations, remote sensing devices, and on-ground sensors. The acquired data is processed using advanced data analytics and

machine learning algorithms to find patterns and trends that may suggest pest or disease outbreaks.

Temperature, humidity, crop growth stage, and historical data are used to determine risk levels. When a substantial danger is discovered, notifications are created and distributed via multiple channels to farmers, extension services, or agricultural authorities. Decision support components are frequently provided, offering advice on relevant actions such as pest control techniques or planting calendar alterations.

Early warning systems enable farmers to respond to possible hazards in real time, decreasing pest and disease transmission and crop losses. For regulating the temperature within a greenhouse, a sizable number of control algorithms have been applied. We can list fundamental controls like Proportional-Integral-Derive (PID) control, fuzzy logic controls, neural network controls, robust controls, non-linear type controls, and hybrid controls [15]. They also assist farmers in using pesticides and herbicides more wisely, lowering environmental effect and productivity costs. Early identification and control guarantee that food production stays steady, lowering the risk of hunger and malnutrition.

These systems encourage sustainable agriculture by improving pest and disease control, which is in line with environmentally friendly farming practices. Farmers may make more informed judgments with data-driven decision-making, which improves overall farm management. Early warning systems encounter difficulties in terms of data quality, infrastructure, and accessibility, particularly in poor countries. Future developments may include the use of artificial intelligence and the growth of sensor networks to increase data quality and coverage.

2.4 IOT SOLUTIONS FOR POST-HARVEST FOOD QUALITY MONITORING

Food quality monitoring after harvest is critical for food safety, quality preservation, economic effect, and consumer satisfaction. During post-harvest handling, storage, and transportation, up to one third of all food produced is lost or squandered. Foodborne disease can be avoided by ensuring that food items are free of pollutants and pathogens. Monitoring also aids in the preservation of the freshness of food, its nutritional content, and its shelf life, hence decreasing spoilage. Reducing losses may boost farmer and food industry profitability, whilst fulfilling consumer expectations is critical for brand reputation and customer loyalty.

Sensors used in IoT systems for post-harvest food quality monitoring provide data on temperature, humidity, gas concentrations, and physical conditions. Storage facilities, transit vehicles, and packaging all have these sensors carefully positioned. Data is delivered through wireless networks to a centralized platform for analysis, which is frequently cloud-based. Advanced data analytics such as machine learning algorithms detect trends and abnormalities in order to forecast probable quality concerns or spoiling incidents.

IoT technologies provide real-time monitoring by allowing stakeholders to access data and get warnings via web-based dashboards or mobile apps. Alerts and messages are provided when established quality criteria are surpassed or abnormalities are found, allowing for rapid actions. Decision support systems routinely give recommendations to address quality issues, such as modifying storage conditions or rerouting transportation routes.

IoT solutions reduce food waste by continuously monitoring environmental conditions and product quality, improve food safety by detecting issues early, optimize resource use by controlling storage conditions, improve supply chain efficiency through real-time data, and enable stakeholders to make data-driven decisions [16]. These advantages help to create a more efficient and sustainable food system by minimizing food deterioration and waste.

2.4.1 Smart Storage and Transportation Systems

Assess your unique crops or goods and their requirements, such as temperature, humidity, and ventilation, to optimize your storage system. Choose a site that is convenient for transit, close to markets or processing facilities, and has the required infrastructure for smart storage systems. Install temperature, humidity, gas, and video sensors to collect data from various regions of the storage space. Using an appropriate platform or software, set up an IoT network to link sensors to a central system. To make appropriate modifications, monitor environmental parameters such as temperature, humidity, and gas levels, install automatic temperature control systems, such as heaters, coolers, humidifiers, and dehumidifiers.

Analyze sensor data to find patterns, abnormalities, and trends. Create algorithms to predict possible problems. Ensure remote access to the data and controls of the storage system for monitoring and adjustment. Integrate pest control solutions such as traps, screens, or automated approaches into your system. Schedule periodic sensor and control system maintenance and calibration to ensure accurate readings and efficient operation. Train employees on data interpretation and maintenance, and keep complete records of environmental conditions, maintenance operations, and storage occurrences. Implement quality assurance tests and inspections to guarantee that the system produces the intended outcomes.

Gather input from your staff on a regular basis and make adjustments based on their views and experiences. Since your firm expands, consider scalability as you may need to extend or adjust your smart storage systems to accommodate higher amounts of produce.

2.4.1.1 IoT-enabled Cold Chain Management

IoT-enabled cold chain management is a complex system that monitors and manages temperature-sensitive supply chains in industries such as food and pharmaceuticals, assuring safe transit and storage of perishable items. Temperature, humidity, and other environmental conditions are monitored in real time by IoT sensors strategically positioned throughout the cold chain. Data is sent via IoT networks to a central monitoring system or cloud-based platform, allowing stakeholders to view it from any location. The incoming data is processed by advanced analytics and machine learning algorithms, which discover abnormalities, patterns, and possible concerns. While the adoption of the so-called Industry 4.0 idea has already made industrial production processes smarter and more autonomous, the use of the IoT, Artificial Intelligence (AI), robots, and big data in agriculture is more recent. The capacity of IoT technology to provide wireless connectivity in rural regions and manage smart items in fields indicates significant improvement potential for the agricultural industry, or so-called smart agriculture [17 –19]. When anomalies or concerns are found, automatic alerts and notifications are generated, which may be delivered to appropriate parties through email, SMS, or mobile applications.

IoT-enabled devices may also modify ambient conditions in real-time, such as triggering cooling mechanisms if the temperature of a refrigerated vehicle climbs over a preset threshold. All data is securely kept for future analysis and compliance needs, allowing stakeholders to develop reports and access past data to better improve the cold chain.

Quality assurance, cost savings, compliance, real-time visibility, predictive maintenance, optimized routes, prolonged shelf life, decreased environmental impact, and enhanced customer satisfaction are all advantages of IoT-enabled systems in the supply chain. Temperature and humidity levels are maintained in these systems to assure the safety and quality of perishable commodities, reducing spoilage and waste. They also aid in meeting regulatory standards by providing precise temperature and environmental information. Furthermore, IoT-enabled systems can predict equipment breakdowns and schedule repair, thus reducing downtime. Overall, IoT-enabled cold chain management helps to improve sustainability and consumer happiness.

2.4.1.2 Smart Packaging for Freshness Preservation

Sensors are used in smart packaging, a novel idea in the food sector, to monitor and increase the shelf life of perishable items. In order to collect information on the environment within the box, it makes use of technologies including temperature, humidity, gas, and biosensors. After this, the gathered data is sent to a central system or cloud-based platform. Real-time data collection, analysis, and processing are made possible by this IoT technology, which enhances farm management and empowers farmers to make better choices [20].

Advanced algorithms and machine learning are utilized to evaluate the obtained data and assess the state and surroundings of the product. When sensor data reveals deviations from the expected conditions, warnings and notifications are provided to customers, merchants, or manufacturers via mobile applications. Some smart packaging options include active ingredients like antimicrobial coatings or oxygen scavengers to improve the shelf life and freshness of the product.

Smart packaging is a technology that enhances freshness preservation by ensuring quality assurance, reducing food waste, enhancing safety, and extending shelf life. It also optimizes inventory management, promotes brand differentiation, and allows real-time monitoring. This technology minimizes spoilage and food waste, promotes trust in the supply chain, and alerts consumers to potential safety risks. It also extends the product's shelf life and enables rapid response to potential issues.

2.5 BENEFITS AND IMPACT OF IOT IN FOOD QUALITY ASSURANCE

The Internet of Things (IoT) has had a substantial influence on food quality assurance throughout the supply chain, from manufacturing to consumption. IoT sensors can monitor environmental conditions in real time, ensuring that food is kept and delivered at acceptable temperatures, lowering the danger of bacterial development and foodborne diseases. Several studies [21] [22] have presented various frameworks to investigate the feasibility of using blockchain. They may also detect changes in food quality, such as ripeness or rotting, enabling the prompt removal of tainted foodstuffs, decreasing food waste, and enhancing overall product quality. Food product tracing along the supply chain enables customers and regulators to track the origin of food items, making it simpler to identify and recall products in the event of contamination or quality concerns. IoT may also be utilized for predictive maintenance and monitoring of food production equipment and machinery.

Real-time data from the IoT devices gives insights into the efficiency of the supply chain, allowing routes to be improved, transportation costs to be minimized, and fresh items to be delivered on time. IoT sensors may assist with inventory management by monitoring product amounts and expiration dates, eliminating overstocking and understocking concerns, and reducing food waste. Smart packaging with IoT characteristics provides consumers with information on product freshness, source, and nutritional content, thereby increasing trust and loyalty. IoT also contributes to food waste reduction by monitoring environmental conditions and product quality, automating compliance monitoring and reporting, and resulting in cost savings.

Another advantage of IoT is data-driven decision-making, since gathered data gives useful insights for enhancing food quality, safety, and overall supply chain performance. The IoT helps to make food systems more sustainable by decreasing food waste, streamlining supply chains, preserving resources, and lowering greenhouse gas emissions.

2.5.1 Improved Food Safety and Reduced Contamination Risks

Implementing Internet of Things (IoT) technology in the food business provides several advantages, including increased food safety and decreased contamination concerns. Critical elements such as temperature, humidity, and air quality are continually monitored by IoT sensors in food storage facilities, transportation vehicles, and processing plants, ensuring ideal conditions for food storage and transit. Early warning systems detect departures from safe conditions and provide warnings when temperatures or humidity levels go outside of permissible ranges, allowing for prompt remedial action to prevent deterioration and contamination.

Data recording enables IoT devices to continuously capture and document temperature and ambient data, giving a full history of food storage conditions. Remote monitoring ensures consistency in safety standards by enabling consistent safety standards across various sites or facilities. IoT enables food product end-to-end traceability, enabling prompt detection and recall in the event of contamination. Compliance monitoring automates adherence to food safety requirements while also encouraging adherence to cleanliness practices. IoT technology helps HACCP systems by delivering real-time data on important control points, enabling proactive steps when circumstances stray from acceptable limits. Human touchpoints are reduced in automated packing and sorting procedures, putting pollutants into the supply chain.

Smart packaging powered by the IoT lets consumers comprehend product freshness, origin, and safety, hence increasing food safety awareness. Visibility of the supply chain aids in identifying contamination sources and implementing corrective actions. However, because internet connectivity is limited in many locations, smart agricultural systems are still in their infancy. Several communication protocols with varied prices, coverage, power consumption, and reliability have been suggested. Despite these advantages, smart agricultural system application remains restricted [23]. Large volumes of data are generated through data analytics, allowing machine learning algorithms to estimate contamination hazards based on prior data, allowing for preventive interventions. The food sector can considerably improve food safety, minimize contamination risks, and respond to concerns more efficiently by utilizing IoT technology, protecting the reputations and financial stability of customers and food businesses.

2.6 ADDRESSING CHALLENGES AND CONCERNS IN IOT IMPLEMENTATION

There are multiple advantages to using IoT (Internet of Things) technologies in various fields, such as food quality assurance and agriculture. It does, however, bring a number of issues and concerns that must be addressed in order for effective implementation and best performance. Because IoT devices collect huge volumes of data, data security and privacy are critical [24]. To prevent unwanted access, data must be encrypted during transmission and storage. User privacy is likewise vital, as are clear policies and permission processes. Interoperability is another issue since IoT systems sometimes include devices from various vendors. By developing industry-wide standards for IoT communication protocols, standardization can help increase interoperability.

Because of the expanding usage of IoT in smart homes, there is a greater demand in scalability, network capacity planning, and effective data management and analytics solutions. Downtime can result in data loss, decreased production, and financial losses; hence, redundancy can help to reduce failures. The research is mostly concerned with energy management via electrical unit control [25–29].

Real-time processing is also an issue, and IoT devices must be power efficient. With IoT installations conforming to data protection and privacy requirements like GDPR and HIPAA, regulatory compliance is critical. Cost control is a continual challenge, with infrastructure expenses and total cost of ownership being significant considerations. Regular evaluations and firmware upgrades are also required to address security risks. Data bias and job displacement are two ethical considerations. To address these issues, parties such as technology suppliers, governments, and end users must work together.

2.6.1 Data Security and Privacy Considerations

Data security and privacy are critical to the adoption of IoT systems, particularly in food quality assurance and agriculture applications [24]. Data encryption, authentication, and access control are important considerations, as are data minimization, consent and transparency, data lifecycle management, secure device management, incident response, regulatory compliance, security by design, third-party services, user education, vendor security assessment, data ownership, and continuous monitoring.

Data encryption guarantees that sent data is encrypted using secure methods, while data at rest prevents theft or unauthorized access to stored data. Authentication procedures validate the identities of people, devices, and apps that access IoT data, whereas access restrictions limit data access to just authorized persons or systems.

Data minimization entails gathering just the information required for the intended purpose while protecting individual identities using techniques such as anonymization and pseudonymization. Users must be informed about data collection procedures and have control over their data in order to provide their consent. Establishing rules and processes for the secure processing of data throughout its lifespan, as well as routinely evaluating and auditing data handling methods to guarantee compliance with privacy requirements, are all part of data lifecycle management. Secure device management methods, including IoT device provisioning and onboarding, are also critical.

Incident response strategies should be designed to address data breaches or security events as soon as possible, and regulatory compliance should be maintained by remaining up-to-date on regulatory developments. Continuous monitoring and auditing of IoT systems may help discover and respond proactively to security and privacy risks. By addressing these aspects, enterprises may increase user trust, reduce data breaches, and maintain compliance with applicable legislation, eventually improving the viability and sustainability of their IoT operations.

2.7 CASE STUDIES AND SUCCESS STORIES

2.7.1 Real-world Examples of IoT Implementation in Food Quality Monitoring

John Deere's Precision Agriculture (PA) technology suite aids farmers in efficient operations by using sensors, GPS, and data collection tools to gather information about soil conditions, crop health, and weather, enabling informed decisions about planting, fertilizing, spraying, and harvesting.

John Deere's PA technologies include:

- AutoTrac is a GPS-based guiding device that assists farmers in driving their equipment straight. This can increase the precision of planting and other activities while also lowering fuel usage.
- Farmers can use Variable Rate Application (VRA) systems to administer inputs like seed, fertilizer, and pesticides at varied rates based on the needs of the land. This can assist to enhance input efficiency and cut expenditures.
- Crop scouting systems acquire data regarding crop health, such as moisture content, nitrogen levels, and disease pressure, using sensors. This data may be used to identify fields that require care, such as irrigation or insect management.

- Farmers may use data management systems to collect, store, and evaluate data from their PA activities. This information may be utilized to track the effectiveness of their operations and make educated decisions about future procedures.

John Deere's PA technologies assist farmers in increasing yields, lowering input costs, and being more sustainable. Farmers that use John Deere PA technology are expected to save 1.3 billion gallons of gasoline, 1.2 million tons of fertilizer, and 1.1 million tons of chemicals by 2020.

Since the introduction of its GreenStar Precision Farming System in 1996, John Deere, a pioneer in precision agriculture, has invested in PA technology research and development. The firm intends to assist farmers in implementing these technologies to improve their operations. Choumert discovered that in industrialized economies, satellite technology enables managers to monitor land resources, send real-time data to information fusion systems, and make real-time choices, allowing for the overall planning of big regional agriculture [30].

Here are some of the advantages of adopting Precision Agriculture from John Deere:

- Yield Increases: PA can help farmers enhance crop yields by up to 10%.
- Reduced Input Costs: PA can help farmers save up to 15% on input expenses such as seed, fertilizer, and pesticides.
- Improved Crop Quality: By lowering stress and pests, PA can assist farmers in improving crop quality.
- PA may assist farms in being more sustainable by lowering their environmental impact.

John Deere's Precision Agriculture technologies can enhance farmer productivity, reduce costs, and promote sustainability.

2.8 CONCLUDING REMARKS

This chapter concludes by emphasizing the transformative potential of IoT technologies in revolutionizing food quality monitoring throughout the entire supply chain. By leveraging AI and the IoT in smart agriculture, stakeholders can make data-driven decisions, optimize production processes, and ensure safer and higher-quality food products for consumers. However, it is essential to address the challenges and concerns surrounding IoT implementation to achieve successful integration in the food industry. The chapter will also outline future prospects and potential research directions to continue advancing the use of AI and IoT in smart agriculture.

REFERENCES

1. Suresh Babu, C. V. (2023a). *IoT and its Applications*. Anniyappa Publication.
2. Tsang, Y. P., Choy, K. L., Wu, C. H., Ho, G. T. S., & Lam, H. Y. (2019). Blockchain-driven IoT for food traceability with an integrated consensus mechanism. *IEEE Access: Practical Innovations, Open Solutions*, 7, 129000–129017. doi: 10.1109/access.2019.2940227.
3. Suresh Babu, C. V., Akshayah, N. S., A., & Janapriyan, R. (2023). IoT-based smart accident detection and alert system. In P. Swarnalatha & S. Prabu (Eds.), *Handbook of Research on Deep Learning Techniques for Cloud-Based Industrial IoT* (pp. 322–337). IGI Global. doi: 10.4018/978-1-6684-8098-4.ch019.
4. Suresh Babu, C. V. (2022). *Artificial Intelligence and Expert Systems*. Anniyappa Publication.
5. Kshetri, N. (2017). Can blockchain strengthen the internet of things? *IT Professional*, 19(4), 68–72. doi: 10.1109/mitp.2017.3051335.
6. Khan, M. A., & Salah, K. (2018). IoT security, blockchain solutions, and open challenges. *Future Generations Computer Systems: FGCS*, 82, 395–411. doi: 10.1016/j.future.2017.11.022.
7. De Haro-Olmo, F. J., Varela-Vaca, Á. J., & Álvarez-Bermejo, J. A. (2020). Blockchain from the perspective of privacy and anonymisation, A systematic literature review. *Sensors (Basel, Switzerland)*, 20(24), 7171. doi: 10.3390/s20247171.
8. Reyna, A., Martín, C., Chen, J., Soler, E., & Díaz, M. (2018). On blockchain and its integration with IoT. *Challenges and Opportunities. Future Generations Computer Systems: FGCS*, 88, 173–190. doi: 10.1016/j.future.2018.05.046.
9. Lee, S., Bae, M., & Kim, H. (2017). Future of IoT networks: A survey. *Applied Sciences (Basel, Switzerland)*, 7(10), 1072. doi: 10.3390/app7101072.
10. Zambon, I., Cecchini, M., Egidi, G., Saporito, M. G., & Colantoni, A. (2019). Revolution 4.0: Industry vs. Agriculture in a future development for SMEs. *Processes (Basel, Switzerland)*, 7(1), 36. doi: 10.3390/pr7010036.
11. Suresh Babu, C. V., Yadavamuthiah, K., Abirami, S., & Kowsika, S. (2023). Artificial Intelligence in wastewater management. In *Artificial Intelligence Applications in Water Treatment and Water Resource Management* (pp. 25–39). doi: 10.4018/978-1-6684-6791-6.ch002.
12. Huang, D., Zhao, D., Wei, L., Wang, Z., & Du, Y. (2015). Modeling and analysis in marine big data: Advances and challenges. *Mathematical Problems in Engineering*, 2015, 1–13. doi: 10.1155/2015/384742.
13. T. Blaschke, S. Lang, and G. Hay. (2008). *Object-Based Image Analysis, Spatial Concepts for Knowledge-Driven Remote Sensing Applications*. Springer Science & Business Media. doi: 10.1007/978-3-540-77058-9.
14. Suresh Babu, C. V., Mahalashmi, J., Vidhya, A., Nila Devagi, S., & Bowshith, G. (2023). Save soil through machine learning. In Maki K. Habib (Ed.) *Global Perspectives on Robotics and Autonomous Systems: Development*

and Applications (pp. 123–140). The American University in Cairo, Egypt. doi: 10.4018/978-1-6684-7791-5.ch016.

15. Behrooz, F., Mariun, N., Marhaban, M., Mohd Radzi, M., & Ramli, A. (2018). Review of control techniques for HVAC systems—nonlinearity approaches based on fuzzy cognitive maps. *Energies*, 11(3), 495. doi: 10.3390/en11030495.
16. Suresh Babu, C. V., & Das, S. (2023). Impact of blockchain technology on the stock market. In K. Mehta, R. Sharma, & P. Yu (Eds.), *Revolutionizing Financial Services and Markets Through FinTech and Blockchain* (pp. 44–59). IGI Global. doi: 10.4018/978-1-6684-8624-5.ch004.
17. Mekala, M. S., & Viswanathan, P. (2017, August). A survey: Smart agriculture IoT with cloud computing. In *2017 International Conference on Microelectronic Devices, Circuits and Systems (ICMDCS). Presented at the 2017 International conference on Microelectronic Devices, Circuits and Systems (ICMDCS), Vellore*. doi: 10.1109/icmdcs.2017.8211551.
18. Liu, Y., Ma, X., Shu, L., Hancke, G. P., & Abu-Mahfouz, A. M. (2021). From industry 4.0 to agriculture 4.0, Current status, enabling technologies, and research challenges. *IEEE Transactions on Industrial Informatics*, 17(6), 4322–4334. doi: 10.1109/tii.2020.3003910.
19. Zhai, Z., Martínez, J. F., Beltran, V., & Martínez, N. L. (2020). Decision support systems for agriculture 4.0, Survey and challenges. *Computers and Electronics in Agriculture*, 170(105256), 105256. doi: 10.1016/j.compag.2020.105256.
20. Tekinerdogan, B. (2018). Architecture design approach for IoT-based farm management information systems. *Precision Agricultural*, 20(5), 926–958.
21. Lo, S. K., Xu, X., Chiam, Y. K., & Lu, Q. (2017, November). Evaluating suitability of applying blockchain. *2017 22nd International Conference on Engineering of Complex Computer Systems (ICECCS)*. Presented at the 2017 22nd International Conference on Engineering of Complex Computer Systems (ICECCS), Fukuoka. doi: 10.1109/iceccs.2017.26.
22. Xu, X., Lu, Q., Liu, Y., Zhu, L., Yao, H., & Vasilakos, A. V. (2019). Designing blockchain-based applications a case study for imported product traceability. *Future 4Generation Computer Systems*, 92, 399–406. doi: 10.1016/j.future.2018.10.010.
23. Avşar, E., & Mowla, M. N. (2022). Wireless communication protocols in smart agriculture, A review on applications, challenges and future trends. *Ad Hoc Networks*, 136(102982), 102982. doi: 10.1016/j.adhoc.2022.102982.
24. Suresh Babu, C. V., & Yadavamuthiah, K. (2023). Precision agriculture and farming using cyber-physical systems: A systematic study. In *Contemporary Developments in Agricultural Cyber-Physical Systems* (pp. 10–29). doi: 10.4018/978-1-6684-7879-0.ch010.
25. Li, W., Logenthiran, T., Phan, V.-T., & Woo, W. L. (2019). A novel smart energy theft system (SETS) for IoT-based smart home. *IEEE Internet of Things Journal*, 6(3), 5531–5539. doi: 10.1109/jiot.2019.2903281.
26. Bouazza, K. E., & Deabes, W. (2019). Smart Petri Nets temperature control framework for reducing building energy consumption. *Sensors (Basel, Switzerland)*, 19(11), 2441. doi: 10.3390/s19112441.

27. Tsai, K.-L., Leu, F.-Y., & You, I. (2016). Residence energy control system based on wireless smart socket and IoT. *IEEE Access, Practical Innovations, Open Solutions*, 4, 2885–2894. doi: 10.1109/access.2016.2574199.
28. Tom, R. J., Sankaranarayanan, S., & Rodrigues, J. J. P. C. (2019). Smart energy management and demand reduction by consumers and utilities in an IoT-fog-based power distribution system. *IEEE Internet of Things Journal*, 6(5), 7386–7394. doi: 10.1109/jiot.2019.2894326.
29. Bedi, G., Venayagamoorthy, G. K., Singh, R., Brooks, R. R., & Wang, K.-C. (2018). Review of IoT in electric power and energy systems. *IEEE Internet of Things Journal*, 5(2), 847–870. doi: 10.1109/jiot.2018.2802704.
30. Choumert, J., & Phélinas, P. (2015). Determinants of agricultural land values in Argentina. *Ecological Economics, The Journal of the International Society for Ecological Economics*, 110, 134–140. doi: 10.1016/j.ecolecon.2014.12.024.

Chapter 3

Leveraging Machine Learning for Weather Condition Prediction and Enhanced Crop Production in Agriculture

C.V. Suresh Babu and A. Auroshaa

3.1 INTRODUCTION

The agricultural sector is the cornerstone of human civilization, providing sustenance and nourishment for communities across the globe. However, the ever-changing and unpredictable nature of weather patterns poses significant challenges to agricultural productivity and food security [1]. In recent years, the integration of machine learning techniques has emerged as a transformative solution, revolutionizing the way we understand, predict, and adapt to weather conditions in agriculture. Weather conditions, including temperature, precipitation, humidity, and wind patterns, play a pivotal role in determining crop health, growth, and yield [2]. Traditional methods of weather forecasting, whilst valuable, often fall short of providing timely and accurate predictions necessary for effective agricultural planning. This is where machine learning steps in, harnessing the power of data-driven insights and algorithms to decipher complex weather patterns and make precise predictions. In this era of digital innovation, a wealth of data is generated through satellite imagery, weather sensors, and historical records [3]. Machine learning algorithms excel at extracting meaningful patterns from this vast data ocean, enabling us to create predictive models that can anticipate weather fluctuations with unprecedented accuracy. By leveraging these models, farmers and agricultural stakeholders can make informed decisions, ranging from selecting optimal planting times and crop varieties to managing irrigation and pest control strategies. One of the remarkable advancements facilitated by machine learning is the development of personalized weather forecasting systems [4]. These systems can provide localized weather predictions tailored to specific agricultural plots, accounting for microclimates and regional variations. Such precision empowers farmers to adopt site-specific strategies, mitigating risks associated with adverse weather events and optimizing resource utilization. Moreover, the marriage of machine learning with agricultural practices extends beyond mere weather prediction. These technologies are instrumental in enhancing crop production through the implementation of smart farming techniques

DOI: 10.1201/9781003676102-3

[5]. By integrating real-time weather data into automated systems, tasks such as irrigation, fertilization, and pest management can be optimized, minimizing waste and maximizing yields. The result is a more sustainable and efficient approach to agriculture that not only ensures food security but also promotes environmental stewardship. As we delve deeper into the realms of machine learning-driven weather prediction and its applications in agriculture, this paper aims to explore the diverse methodologies, challenges, and success stories that define this evolving landscape. By examining real-world case studies and cutting-edge research, we will unravel the potential of machine learning to revolutionize weather forecasting and crop production, paving the way for a more resilient and prosperous agricultural future.

Agriculture, as the cornerstone of human sustenance, has always been intricately entwined with the capricious forces of nature. Weather conditions wield significant influence over crop growth, health, and yield, shaping the course of agricultural livelihoods and global food security. However, with the advent of technological advancements, the integration of machine learning has ushered in a new era of precision, innovation, and resilience in agricultural practices. The challenge of predicting weather conditions with accuracy has long been a pursuit of meteorologists and agricultural scientists. In parallel, the evolution of machine learning has unlocked the potential to decipher the intricate tapestry of weather phenomena by processing vast amounts of data and recognizing patterns that may elude traditional methods. Machine learning algorithms, guided by historical weather data, satellite imagery, and other relevant variables, have demonstrated a remarkable capacity to uncover hidden correlations, make accurate predictions, and offer insights that can revolutionize agricultural decision-making [6]. From precise irrigation scheduling based on soil moisture predictions to strategic planning decisions informed by anticipated temperature trends, these innovations empower agriculturists to make informed choices that drive efficiency and productivity while minimizing resource waste.

Against this backdrop, this study delves into the synergy between machine learning, weather condition prediction, and enhanced crop production. Machine learning's ability to analyse historical data and model future scenarios offers a potent tool for anticipating and mitigating the impacts of a changing climate on agriculture.

3.1.1 Objectives of this Chapter

Here, we delve into the intricate relationship between weather conditions and crop growth, highlighting how different meteorological variables impact various stages of plant development. We aim to establish a solid foundation for appreciating the significance of accurate weather predictions in agricultural planning. We:

- Provide an in-depth exploration of various machine learning algorithms and methodologies suitable for weather prediction and crop management. We detail how these techniques analyse historical and real-time data to generate accurate forecasts and valuable insights [7].
- Present the process of building robust machine learning models for weather forecasting.
- Demonstrate how machine learning algorithms can provide hyper-localized weather predictions, enabling farmers to make informed decisions and adapt farming practices accordingly.
- Explore the integration of machine learning into crop management practices, elucidating how real-time weather data can optimize irrigation, fertilization, and pest control strategies. We discuss the potential for resource efficiency, reduced environmental impact, and improved crop yields through data-driven decision-making [8].

3.2 SIGNIFICANCE OF WEATHER PREDICTION IN AGRICULTURE

Weather prediction has emerged as a pivotal factor in modern agriculture, acting as a critical bridge between scientific knowledge and practical decision-making for farmers and stakeholders. The significance of weather prediction in agriculture, amplified by machine learning, can be understood through the following key points:

- Machine learning models provide insights into soil moisture levels, humidity, and temperature trends. This information empowers farmers to allocate resources such as water and fertilizers more efficiently, thus reducing waste and promoting sustainable resource management [9].
- Machine learning-enabled weather prediction assists farmers in adapting to changing climate patterns. As climate change brings about shifts in temperature and precipitation, these predictive models allow for proactive adjustments to planting and harvesting seasons.
- The integration of machine learning and weather prediction contributes to the advancement of scientific understanding of weather patterns and their impacts on agriculture. This knowledge can inform policy decisions, research initiatives, and sustainable development goals [10].

In essence, the significance of weather prediction in agriculture, coupled with machine learning capabilities, lies in its capacity to empower farmers with actionable insights that drive efficiency, productivity, and sustainability. As agricultural systems navigate the challenges posed by a changing climate and a growing global population, accurate and data-driven weather predictions emerge as an indispensable tool for shaping the future of farming.

3.3 WEATHER CONDITION PREDICTION USING MACHINE LEARNING

Weather has a direct impact on crop growth, yield, and total agricultural production in agriculture. Accurate weather forecasting is critical for farmers and agricultural stakeholders to make educated decisions regarding planting, irrigation, and harvesting. This article investigates the use of machine learning in weather prediction and its implications for agriculture.

3.3.1 Overview of Machine Learning Algorithms

3.3.1.1 Supervised Learning Algorithms

Linear Regression: Used to predict continuous variables, such as temperature or precipitation amounts [11].

Support Vector Machines (SVM): Can classify weather conditions, such as clear, cloudy, or rainy, based on historical data [12].

Random Forest: Effective for predicting crop yields based on multiple environmental factors like temperature, humidity, and soil quality [12].

3.3.1.2 Deep Learning Algorithms

Convolutional Neural Networks (CNNs): Used for analyzing weather images, such as satellite images, to predict weather patterns, cloud cover, and storm tracking [13].

Long Short-Term Memory (LSTM): Effective for capturing temporal dependencies in weather data and predicting weather conditions over extended periods. [14]

3.2.1.3 Unsupervised Learning

Clustering Algorithms: Grouping similar weather patterns can help identify recurring climate conditions, which can be valuable for predicting agricultural outcomes [15].

3.2.1.4 Reinforcement Learning

Optimizing Irrigation: Reinforcement learning can be used to develop irrigation strategies that maximize crop yield while conserving water resources, and adapting to changing weather conditions [15].

3.2.1.5 Anomaly Detection

Identifying Extreme Weather Events: Anomaly detection algorithms can help identify unusual weather patterns that may indicate the onset of extreme events like droughts, floods, or storms [16].

3.3.2 Data Collection and Preprocessing

3.3.2.1 Data Collection

Weather Data Sources: Collect historical weather data from various sources, including meteorological stations, weather APIs, satellites, and remote sensing technologies.
Crop Data Sources: Gather crop-specific data, including planting and harvesting dates, crop types, and yield data.
Remote Sensing: Employs satellite and aerial imagery for monitoring crop health, detecting anomalies, and assessing vegetation indices.
IoT Devices: Can be deployed as, for example, weather stations and soil moisture sensors across fields for real-time data collection. [17]

3.3.2.2 Data Preprocessing

Data Cleaning: Missing data can be handled by imputing values using methods like mean, median, or interpolation.
Data Integration: Weather, crop, and soil data can be combined into a unified dataset for analysis.
Feature Engineering: New features can be created based on domain knowledge, such as Growing Degree Days (GDD), chill hours, or crop-specific variables.
Normalization and Scaling: Numerical features can be normalized to have a consistent scale (for example, using Min-Max scaling or Z-score normalization).
Scaling Features: Features can be scaled to prevent certain variables from dominating the learning process.
Encoding Categorical Variables: Categorical variables (for example, crop types) can be converted into numerical representations using techniques like one-hot encoding or label encoding.

3.3.3 Feature Selection and Engineering

3.3.3.1 Feature Selection

Correlation Analysis: Correlations can be calculated between weather variables (for example, temperature, precipitation) and crop yield or other relevant outcomes. Select features with the highest correlations as they may have the most significant impact on crop production.

Feature Importance from Models: Tree-based models like Random Forest or Gradient Boosting can be used to assess feature importance. Select the top-ranked features according to their contribution to model accuracy.

Recursive Feature Elimination (RFE): RFE techniques can be applied to iteratively remove the least important features and retrain models until the desired number of features is achieved.

3.3.3.2 Feature Engineering

Temporal Features: Features can be created that capture temporal patterns, such as day of the year, seasonal indicators, or time since the last rainfall. Calculate rolling statistics like moving averages to smooth out noise in time series data.

Spatial Features: Features can be generated that account for geographical variations, such as elevation, proximity to water bodies, or soil type. Incorporate data from GPS and GIS sources to create spatial indicators.

Weather Indices: Weather indices relevant to agriculture, such as Growing Degree Days (GDD), heat stress indices, or chill hours can be calculated. These indices can provide insights into crop growth stages and suitability for specific crops.

3.3.3.3 Model Training and Evaluation

Machine learning can be used to predict weather conditions by training models on historical weather data. The type of machine learning model used will depend on the specific application, but some common techniques include:

- Support Vector Machines (SVMs): SVMs are a type of supervised learning algorithm that can be used to classify or regress data. They are known for their accuracy and robustness [17].
- Artificial Neural Networks (ANNs): ANNs are a type of unsupervised learning algorithm that can be used to learn complex relationships between data. They are often used for image recognition and natural language processing, but they can also be used for weather prediction [18].
- Time Series Analysis: Time series analysis is a statistical technique that can be used to analyze data that is collected over time. It can be used to predict future values of a time series, such as the weather [18].

The accuracy of a machine learning model for weather prediction will depend on the quality of the training data, the complexity of the model, and the specific application. In general, more data will lead to more accurate predictions. However, it is important to note that no machine learning model is perfect, and there will always be some uncertainty in weather predictions.

- Mean Absolute Error (MAE): The MAE is the average of the absolute difference between the predicted and actual values.
- Mean Squared Error (MSE): The MSE is the average of the squared differences between the predicted and actual values.
- Root Mean Squared Error (RMSE): The RMSE is the square root of the MSE.

3.3.5 Challenges in Weather Prediction

There are a number of challenges in leveraging machine learning for weather condition prediction and enhanced crop production in agriculture. Some of these challenges include:

- Blockchain technology: This addresses inefficiencies and risks in traditional stock market infrastructure by providing a decentralized and transparent platform for investors and issuers to exchange assets without middlemen, addressing the need for historical weather data for machine learning models [19].
- Shortage of computational power: Training machine learning models can require a lot of computational power, which can be a challenge for farmers and other agricultural stakeholders who may not have access to powerful computers [20].
- Interpretability: It can be difficult to interpret the predictions made by machine learning models, which can make it difficult to understand why the model made a particular prediction. This can be a challenge for farmers who need to understand the basis for the model's predictions to make informed decisions about crop production.

3.4 SUPERVISED MACHINE LEARNING FOR WEATHER FORECASTING

Supervised machine learning is effective for weather forecasting, and it is becoming increasingly popular as more data becomes available. However, it is important to note that no machine-learning model is perfect, and there will always be some uncertainty in weather predictions.

Here are some of the benefits of using supervised machine learning for weather forecasting:

- It can be used to predict weather conditions over a wide range of time scales, from a few hours to a few weeks.
- It can be used to predict weather conditions at a variety of spatial scales, from a single location to a large region.

3.4.1 Support Vector Machines (SVM)

In SVM classification, the goal is to find a hyperplane that separates the two classes of data points with the maximum margin. The margin is the distance between the hyperplane and the closest data points of each class. The SVM algorithm finds the hyperplane that maximizes the margin, which helps to improve the generalization performance of the model [21].

In SVM regression, the goal is to find a function that maps the input data to the output data with the minimum error. The SVM algorithm finds the function that minimizes the error, subject to the constraint that the function is as flat as possible. This helps to improve the smoothness of the model and prevent overfitting.

SVMs can be used for a variety of tasks, including:

- Classification: SVMs can be used to classify data into two or more categories. For example, SVMs can be used to classify images as spam or ham, or to classify text as positive or negative sentiment.
- Regression: SVMs can be used to predict a continuous value, such as the price of a house or the severity of a disease.

3.4.2 Decision Trees and Random Forests

Decision trees and random forests are both machine learning algorithms that can be used for classification or regression tasks. However, there are some key differences between the two algorithms [22].

Random forests are a type of ensemble learning algorithm that combines multiple decision trees. The random forest algorithm builds each decision tree on a different subset of the data and with a different subset of the features. This helps to reduce the variance of the predictions and improve the overall accuracy of the model.

3.4.3 Neural Networks for Weather Prediction

Neural networks are a type of machine learning algorithm that can be used to learn complex relationships between data. They are often used for image recognition and natural language processing, but they can also be used for weather prediction [23].

Neural networks are inspired by the human brain. They are made up of interconnected nodes, called neurons, that are arranged in layers. The first layer, called the input layer, receives the data. The last layer, called the output layer, produces the predictions. The intermediate layers, called hidden layers, learn to extract features from the data and pass them on to the output layer.

Here are some of the benefits of using neural networks for weather forecasting:

- They can learn complex relationships between weather variables.
- They can be used to predict weather conditions over a wide range of time scales.

3.4.4 Comparative Analysis of Different Algorithms

The choice of which algorithm to use for a comparative analysis will depend on the specific task and the availability of data. For example, if the goal is to predict a continuous value, then SVMs or neural networks might be a good choice. However, if the goal is to classify data into two or more categories, then decision trees or random forests might be a better choice.

The results of a comparative analysis can be used to choose the best algorithm for a particular task. They can also be used to improve the accuracy of an algorithm by identifying the factors that affect its performance [24].

3.5 LEVERAGING HISTORICAL WEATHER DATA FOR PREDICTION

Historical weather data can be used to identify trends in weather patterns. This can be helpful for predicting future weather conditions. For example, if historical data shows that a particular region experiences more rainfall in the summer months, then it is likely that the region will experience more rainfall in the upcoming summer months [25].

Historical weather data can be used to predict extreme weather events, such as hurricanes, tornadoes, and floods. This can help to mitigate the damage caused by these events. For example, if historical data shows that a particular region is prone to hurricanes, then residents of that region can be warned to take precautions before a hurricane strikes.

3.5.1 Data Sources and Repositories

- National Oceanic and Atmospheric Administration (NOAA): The NOAA is a US federal agency that is responsible for collecting and providing weather data. NOAA's National Centers for Environmental Information (NCEI) maintains a comprehensive archive of weather data, including historical weather data, real-time weather data, and satellite data [25].

- European Centre for Medium-Range Weather Forecasts (ECMWF): ECMWF is an international organization that provides weather forecasts and climate data. ECMWF's archive of weather data includes historical weather data, real-time weather data, and model output data.
- Weather Underground: Weather Underground is a private weather service that provides weather data and forecasts. Weather Underground's archive of weather data includes historical weather data, real-time weather data, and station data.

3.5.2 Preparing Historical Weather Data for Training

The first step is to collect the historical weather data. This can be done from a variety of sources, such as the National Oceanic and Atmospheric Administration (NOAA), the European Centre for Medium-Range Weather Forecasts (ECMWF), and NASA.

The data needs to be encoded before it can be used for training. This is because machine learning algorithms can only understand numerical data. The data can be encoded using a variety of techniques, such as one-hot encoding and ordinal encoding [26].

The data needs to be split into a training set and a test set. The training set will be used to train the model, and the test set will be used to evaluate the model.

3.5.3 Handling Missing Data and Anomalies

Missing values can be imputed using a variety of methods, such as the mean, median, or mode. The mean is the average of all the values in the dataset. The median is the middle value in the dataset when all the values are sorted from least to greatest. The mode is the most frequent value in the dataset.

Missing values can be removed from the dataset. However, this can reduce the size of the dataset and may affect the accuracy of the model.

The data can be transformed to reduce the impact of outliers. This can be done by using a technique called normalization. Normalization scales the data so that all the values are between 0 and 1.

3.6 APPLICATIONS OF WEATHER PREDICTION IN AGRICULTURE

Weather prediction can help farmers choose the best time to plant their crops. For example, farmers may want to plant their crops after the last frost has passed, or before the onset of a drought.

Weather prediction can help farmers determine how much water to use to irrigate their crops. For example, farmers may need to irrigate their crops more during a drought, or less during a rainy season [26].

Weather prediction is a powerful tool that can help farmers make informed decisions about their crops. As weather prediction technology continues to improve, even more applications for weather prediction in agriculture will likely be developed.

3.6.1 Integration of Weather Forecasts into Agricultural Practices

Weather forecasts can be used to determine how much water to use to irrigate crops. For example, farmers may need to irrigate their crops more during a drought, or less during a rainy season [27].

Weather forecasts can be used to time pest control applications. For example, farmers may need to apply pesticides before a pest infestation occurs, or after a pest infestation has been detected.

3.6.2 Optimal Irrigation Scheduling Based on Weather Predictions

These methods use simple rules to determine when to irrigate, such as irrigating when the soil moisture is below a certain level.

Decision trees are a type of machine learning algorithm that can be used to create a decision-making flowchart. The flowchart can be used to determine when to irrigate based on a variety of factors, such as the weather forecast, the type of crop, and the soil moisture.

Neural networks are a type of machine learning algorithm that can be used to learn complex relationships between data. Neural networks can be used to create models that can predict when to irrigate based on weather forecasts and other factors [28].

Optimal irrigation scheduling can help to reduce soil erosion by preventing the soil from becoming too dry or too wet [28].

Optimal irrigation scheduling can help to increase profits by reducing the cost of water and improving crop yields.

3.6.3 Precision Fertilization and Crop Management

This technology uses a map of the field to apply different amounts of fertilizer to different areas of the field. This is based on factors such as soil type, crop type, and yield goals [29].

This technology uses GPS to guide tractors and other equipment to the correct location in the field. This ensures that fertilizer is applied accurately and efficiently.

3.6.4 Pest and Disease Management Strategies

Cultural controls: Cultural controls are methods that involve changing the environment to make it less favorable for pests and diseases. This can include practices such as crop rotation, sanitation, and planting resistant varieties [29].

Chemical controls: Chemical controls are methods that use pesticides to kill pests or diseases. This is the most common method of pest and disease control, but it can also be the most harmful to the environment [30].

3.7 CASE STUDIES IN SMART AGRICULTURE

Napa Valley is a world-renowned wine region in California. In recent years, many Napa Valley wineries have begun to adopt precision viticulture techniques. Precision viticulture is a term used to describe the use of technology to improve the efficiency and sustainability of wine production [31].

One of the most common precision viticulture techniques used in Napa Valley is the use of drones [31]. Drones can be used to collect data on a variety of factors, such as crop health, soil moisture, and pest infestation. This data can then be used to make decisions about irrigation, fertilization, and crop protection [31].

Another precision viticulture technique used in Napa Valley is the use of sensors. Sensors can be used to collect data on soil moisture, temperature, and light levels. This data can then be used to optimize irrigation and crop management practices.

Precision viticulture has been shown to improve the efficiency and sustainability of wine production in Napa Valley. For example, one study found that wineries that used precision viticulture techniques were able to increase their yields by 10% while reducing their water usage by 20% [32].

Here are some of the benefits of precision viticulture in Napa Valley:

- Increased yields: Precision viticulture can help to increase yields by improving the efficiency of irrigation and crop management practices [33].
- Reduced water usage: Precision viticulture can help to reduce water usage by improving the efficiency of irrigation practices.

3.7.1 Successful Implementations of Weather Prediction Models

Drones can be employed to collect data on crop health, soil moisture, and pest infestation. This data can then be analysed to make decisions about irrigation, fertilization, and crop protection. For example, drones can be used to monitor the health of grapevines and identify areas that are at risk of pests or diseases. This information can then help to target irrigation and fertilization more precisely, which can aid in improving crop yields and quality.

- Sensors can collect data on soil moisture, temperature, and light levels. This data can then be used to optimize irrigation and crop management practices. For example, sensors can measure the soil moisture in different parts of a vineyard. This information can then be used to irrigate the vineyard more precisely, which can help to save water and improve crop yields.
- Weather prediction models can forecast future weather conditions. This information can then be used to make decisions about crop planting, irrigation, and pest control. For example, weather prediction models can be used to forecast the likelihood of a frost, which can help farmers to decide when to plant their crops.

3.7.2 Real-world Impact on Crop Yields and Farming Practices

Weather prediction models can help farmers to make better decisions about crop planting, irrigation, and pest control. This can lead to increased yields. For example, weather prediction models can be used to forecast the likelihood of a frost, which can help farmers to decide when to plant their crops. This can help to avoid crop losses due to frost [34].

To achieve reduced water usage, weather prediction models can help farmers to optimize their irrigation practices. This can help to reduce water usage and save farmers money. For example, weather prediction models can be used to forecast the amount of rainfall that is expected in a particular area. This information can then be used to adjust irrigation schedules accordingly. Weather prediction models can help farmers to monitor crop health and identify potential problems early on. This can help to improve crop quality by preventing pests and diseases from damaging the crops. For example, weather prediction models can be used to forecast the likelihood of a heat wave, which can stress crops and make them more susceptible to pests and diseases. This information can then be used to take steps to protect the crops, such as irrigating them more frequently.

3.8 CHALLENGES AND FUTURE DIRECTIONS

3.8.1 Challenges

- The availability and quality of weather data can vary greatly, depending on the location and time period. This can make it difficult to train accurate weather prediction models.

- The weather system is a complex system that is influenced by many factors, such as the sun, the atmosphere, and the oceans. This makes it difficult to develop accurate weather prediction models [34].
- The weather system is not stationary, meaning that the statistical properties of the system change over time. This makes it difficult to develop accurate weather prediction models that can be used for long-term forecasting.

3.8.2 Future Directions

- Researchers are constantly developing new machine learning algorithms that can be used to improve the accuracy of weather prediction models.
- Integration of machine learning with other technologies: Researchers are working on integrating machine learning with other technologies, such as sensor networks and drones, to improve the accuracy and efficiency of weather prediction and crop production.

3.8.3 Data Quality and Reliability

Data quality is important because it affects the accuracy of the machine learning models. If the data is not accurate, then the models will not be accurate either. There are a number of factors that can affect data quality, such as:

- Incomplete data: Incomplete data can lead to bias in the models. For example, if a model is trained on data that only includes sunny days, then the model may not be able to predict weather conditions accurately on cloudy days.
- Outliers: Outliers are data points that are significantly different from the rest of the data. Outliers can also lead to bias in the models. For example, if a model is trained on data that includes a few very high rainfall values, then the model may overestimate the amount of rainfall that is expected.
- Noise: Noise is random variation in the data. Noise can also lead to bias in the models. For example, if a model is trained on data that includes a lot of noise, then the model may not be able to accurately predict the underlying trends in the data.

Data reliability is also important because it affects the trust that users have in the machine learning models. If the data is not reliable, then users may not trust the models, even if they are accurate. There are a number of factors that can affect data reliability, such as:

- Data source: The source of the data can affect its reliability. For example, data from a government agency is generally considered to be more reliable than data from a private company [35].
- Data processing methods: The methods used to process the data can also affect its reliability. For example, data that is processed using rigorous statistical methods is generally considered to be more reliable than data that is processed using less rigorous methods.

3.8.4 Overcoming Limitations in Weather Prediction Models

- As mentioned earlier, the availability and quality of weather data can vary greatly, depending on the location and time period. This can make it difficult to train accurate weather prediction models. Researchers are working on improving the collection and processing of weather data to make it more available and accurate.
- Researchers are constantly developing new machine learning algorithms that can be used to improve the accuracy of weather prediction models. These algorithms are often more robust to noise and outliers, and they can handle the complexity of the weather system.
- Researchers are also working on developing more robust machine learning models that can handle the complexity and non-stationarity of the weather system. These models are often trained on larger datasets and they are able to learn the underlying patterns in the data more effectively.
- As mentioned earlier, the agricultural sector is heterogeneous, meaning that different crops and farming practices have different needs. This makes it difficult to develop one-size-fits-all machine learning models for agriculture. Researchers are working on developing machine learning models that are specifically tailored to specific crops and farming practices. This will help to improve the accuracy and applicability of these models.

3.8.5 Integration of IoT and Weather Sensors in Agriculture

- By using the IoT and weather sensors to monitor crop health and manage irrigation and fertilization, farmers can improve crop yields.
- By using the IoT and weather sensors to optimize crop management practices, farmers can reduce costs, such as the cost of water and fertilizer [35].
- By using the IoT and weather sensors to monitor pest and disease infestations, farmers can protect their crops from pests and diseases.

- By using the IoT and weather sensors to reduce water and fertilizer use, farmers can help to make agriculture more sustainable.

3.8.6 Potential for Climate Change Adaptation

- By using IoT and weather sensors to monitor crop health and manage irrigation and fertilization, farmers can improve crop yields. This is because they can optimize their crop management practices to adapt to the changing climate. For example, if the climate is becoming drier, they can use IoT sensors to monitor soil moisture levels and irrigate crops only when necessary.
- By using the IoT and weather sensors to optimize crop management practices, farmers can reduce costs, such as the cost of water and fertilizer. This is because they can use the data collected by the sensors to make more efficient use of these resources. For example, if the climate is becoming warmer, they can use IoT sensors to monitor crop water requirements and apply water only when necessary [35].
- By using the IoT and weather sensors to monitor pest and disease infestations, farmers can protect their crops from pests and diseases. This is because they can use the data collected by the sensors to identify potential problems early on and take action to prevent them from becoming worse. For example, if the climate is becoming warmer, they can use IoT sensors to monitor pest populations and apply pesticides only when necessary.

3.9 CONCLUSION

Machine learning is a powerful tool that can be used to improve weather condition prediction and enhanced crop production in agriculture. By integrating machine learning with other technologies, such as the IoT and weather sensors, farmers can collect real-time data on a variety of factors that can affect crop production. This data can then be used to make informed decisions about crop management, which can help to improve yields, reduce costs, and protect crops from pests and diseases. By using machine learning to predict weather conditions and optimize crop management practices, farmers can improve crop yields. By using machine learning to make more efficient use of resources, such as water and fertilizer, farmers can reduce costs. By using machine learning to monitor pest and disease infestations and apply pesticides only when necessary, farmers can protect their crops from pests and diseases. By using machine learning to reduce water and fertilizer use, farmers can help to make agriculture more sustainable. The integration of machine learning with other technologies is a promising approach to improving weather condition prediction and enhanced crop production in agriculture. By collecting real-time data on a variety of factors that can affect crop production, farmers can make informed decisions about crop management, which can help to improve yields, reduce costs, and protect crops from pests and diseases.

3.9.1 Summary of Key Findings

Machine learning can be used to improve weather condition prediction and enhanced crop production in agriculture. By integrating machine learning with other technologies, such as the IoT and weather sensors, farmers can collect real-time data on a variety of factors that can affect crop production. This data can then be used to make informed decisions about crop management, which can help to improve yields, reduce costs, and protect crops from pests and diseases.

3.9.2 Implications for the Future of Smart Agriculture

- By using machine learning to predict weather conditions and optimize crop management practices, farmers can improve crop yields. This is especially important in the face of climate change, which is making it more difficult to predict and manage weather conditions.
- By using machine learning to make more efficient use of resources, such as water and fertilizer, farmers can reduce costs. This can help to make agriculture more sustainable.
- By using machine learning to monitor pest and disease infestations and apply pesticides only when necessary, farmers can protect their crops from pests and diseases. This can help to reduce the use of pesticides and improve the health of the environment.
- By using machine learning to reduce water and fertilizer use, farmers can help to make agriculture more sustainable. This is important for the future of the planet, as agriculture is a major contributor to climate change.
- Machine learning can help farmers to be more efficient in their operations. For example, machine learning can be used to optimize crop planting and harvesting schedules, which can help to reduce waste and improve yields.
- Machine learning can help farmers to make better decisions about their operations. For example, machine learning can be used to analyze data on crop yields, weather conditions, and market prices to help farmers make decisions about what crops to plant, when to plant them, and how much to sell them for.
- Machine learning can help farmers to increase their profitability. For example, machine learning can be used to identify opportunities to save money on inputs, such as water and fertilizer.

REFERENCES

1. S. V. Prathiksha, N. S. Spandana, S. L. Naveen, U. M. Ashwinkumar, 2023, "Crop Yield Prediction of Cotton Using Optimization Technique", in *2023 International Conference on Smart Systems for Applications in Electrical Sciences (ICSSES)*, pp. 1–6, IEEEXplore, Tumakuru, In, doi: 10.1109/ICSSES58299.2023.10200233.
2. T. Deshmukh, A. Rajawat, S. B. Goyal, J. Kumar, A. Potgantwar, 2023, "Analysis of Machine Learning Technique for Crop Selection and Prediction of Crop Cultivation", in *2023 International Conference on Inventive Computation Technologies (ICICT)*, pp. 298–311, IEEEXplore, Lalitpur, Nepal. doi: 10.1109/ICICT57646.2023.10134371.
3. C. S. Suresh Babu, A. Swapna, D. S. Chowdary, B. S. Vardhan, Imran, 2023, "M. Leaf Disease Detection Using Machine Learning (ML)", in A. Khang (ed.), *Handbook of Research on AI-Equipped IoT Applications in High-Tech Agriculture*. IGI Global, pp. 188–199. doi: 10.4018/978-1-6684-9231-4.ch010.
4. C. V. Suresh Babu, A. Sanjai Das, 2023, "Impact of Blockchain Technology on the Stock Market", in K. Mehta, R. Sharma, P. Yu (eds.), *Revolutionizing Financial Services and Markets Through FinTech and Blockchain*. IGI Global, pp. 44–59. doi: 10.4018/978-1-6684-8624-5.ch004.
5. C. V. Suresh Babu, J. Mahalashmi, A. Vidhya, S. Nila Devagi, G. Bowshith, 2023, "Save Soil Through Machine Learning", in M. Habib (ed.), *Global Perspectives on Robotics and Autonomous Systems: Development and Applications*. IGI Global, pp. 345–362 doi: 10.4018/978-1-6684-7791-5.ch016.
6. G. Sharma, S. K. Dubey, 2023, "Approach towards Weather Prediction Model for Aggrotech", in *2023 Third International Conference on Artificial Intelligence and Smart Energy (ICAIS)*, pp. 1170–1176, IEEEXplore, Coimbatore, India. doi: 10.1109/ICAIS56108.2023.10073904.
7. C. Sagana, M. Sangeetha, S. Savitha, K. Devendran, T. Kavin, K. Kavinsri, P. Mithun, 2023, "Machine Learning-Based Crop Recommendations for Precision Farming to Maximize Crop Yields", in *2023 International Conference on Computer Communication and Informatics (ICCCI)*, pp. 1–5, IEEEXplore, Coimbatore, India. doi: 10.1109/ICCCI56745.2023.10128525.
8. A. Fondaj, M. Hamiti, S. Krrabaj, J. Ajdari, X. Zenuni, 2023, "A Prediction Model of Smart Agriculture Based on IoT Sensor Data: A Systematic Literature Review", in *2023 12th Mediterranean Conference on Embedded Computing (MECO)*, pp. 1–8, IEEEXplore, Budva, Montenegro. doi: 10.1109/MECO58584.2023.10154965.
9. S. J. Rayen, M. S. Jacob, S. Nivetha, L. S. Raveendran, J. S. Premalatha, S. S. Clarinda, 2023, "An IoT Assisted Weather Prediction and Information Monitoring Scheme based on Intensive Learning Strategy", in *2023 International Conference on Advances in Computing, Communication and Applied Informatics (ACCAI)*, pp. 1–10, IEEEXplore, Chennai, India. doi: 10.1109/ACCAI58221.2023.10200067.

10. Y. Ommane, M. A. Rhanbouri, H. Chouikh, M. Jbene, I. Chairi, M. Lachgar, S. Benjelloun, 2023, "Machine Learning Based Recommender Systems for Crop Selection: A Systematic Literature Review", *Machine Intelligence for Smart Applications*, vol. 1105, p. 21, ResearchSquare PrePrint Publisher. doi: 10.1007/978-3-031-37454-8_2.
11. T. Chandrakumar, D. Sakthipriya, S. Devi Mahalakshmi, 2023, "A Study on Usage of Agricultural Engineering Equipment for Various Crops and Yields in South Tamilnadu", *International Journal of Social Ecology and Sustainable Development*, vol. 14, no. 1, p. 1. doi: 10.4018/IJSESD.322014.
12. N. Patel, A. Shah, Y. Soni, N. Solanki, M. Shahu, 2023, "A Survey on Crop Rotation Using Machine Learning and IoT", *IOT with Smart Systems*, vol. 312, p. 57. doi: 10.1007/978-981-19-3575-6_7.
13. M. Shinde, R. Nare, J. Shikalgar, R. Patil, M. Tamboli, 2022, "Plant Growth Recommendation System based on Weather Conditions and Soil Patterns using Machine Learning and Deep Learning Techniques", in *2022 International Conference on Smart Generation Computing, Communication and Networking (SMART GENCON)*, pp. 1–5, IEEExplore. doi: 10.1109/SMARTGENCON56628.2022.10084211.
14. S. Bharti, P. Kaur, P. Singh, C. Madhu, N. Garg, 2022, "Crop Yield Prediction Using CNN-LSTM Model", in *2022 IEEE Conference on Interdisciplinary Approaches in Technology and Management for Social Innovation (IATMSI)*, pp. 1–5, IEEExplore. doi: 10.1109/IATMSI56455.2022.10119308.
15. Saad Bin Manjur Adit, Imtiaz Akber Chowdhury, Wahidul Alam, S M Salam, 2022, "A Predictive Analysis on Weather Parameters Based on Bangladesh Agriculture Using Machine Learning Approach", in *2022 IEEE North Karnataka Subsection Flagship International Conference (NKCon)*, pp. 1–5, IEEExplore. doi: 10.1109/NKCon56289.2022.10127064.
16. J. Mahale, S. Degadwala, D. Vyas, 2022, "Crop Prediction System based on Soil and Weather Characteristics", in *2022 Sixth International Conference on I-SMAC (IoT in Social, Mobile, Analytics and Cloud) (I-SMAC)*, pp. 340–345, IEEExplore. doi: 10.1109/I-SMAC55078.2022.9987366.
17. S. Gaud, R. Sarambale, S. Kale, P. Gunjgur, 2022, "A Recommendation System for Integrated Agriculture Using Neural Networks with Random Forest Algorithm", in *2022 6th International Conference On Computing, Communication, Control And Automation (ICCUBEA*, pp. 1–6, IEEExplore. doi: 10.1109/ICCUBEA54992.2022.10010903.
18. N. Valarmathi, N. Srija, S. Ananthi, Bhuvaneshwari P, Elakkiya T, Nishithraa N, 2022, "Amenity Agriculture App using Random Forest Algorithm", in *2022 International Conference on Sustainable Computing and Data Communication Systems (ICSCDS)*, pp. 522–527, IEEExplore. doi: 10.1109/ICSCDS53736.2022.9760901.
19. M. Aruna Devi, D. Suresh, D. Jeyakumar, D. Swamydoss, M. Lilly Florence, 2022, "Agriculture Crop Selection and Yield Prediction using Machine Learning Algorithms", in *2022 Second International Conference on Artificial Intelligence and Smart Energy (ICAIS)*, pp. 510–517, IEEExplore. doi: 10.30574/wjarr.2022.14.3.0581.

20. A. Ikram, W. Aslam, R. H. H. Aziz, F. Noor, G. A. Mallah, S. Ikram, M. S. Ahmad, A. M. Abdullah, I. Ullah, 2022, "Crop Yield Maximization Using an IoT-Based Smart Decision", *Journal of Sensors*, vol. 2022, p. 1. doi: 10.1155/2022/2022923.
21. K. Jain, N. Choudhary, 2022, "Comparative Analysis of Machine Learning Techniques for Predicting Production Capability of Crop Yield", *International Journal of System Assurance Engineering and Management,* Vol 13, pp 583–593. doi: 10.1007/s13198-021-01543-8.
22. Y. Shahare, V. Gautam, 2022, "Soil Nutrient Assessment and Crop Estimation with Machine Learning Method: A Survey", *Cyber Intelligence and Information Retrieval*, vol. 291, p. 253. doi: 10.32628/IJSRST207433.
23. K. Prakash, K. S. S. Harsha, P. S. Jeevan, L. R. Chandran, T.S. Angel, 2022, "Machine Learning-Based Crop Prediction: A Way Towards Smart Farming", *2022 Seventh International Conference on Parallel, Distributed and Grid Computing (PDGC)*, pp. 623–628, IEEExplore. doi: 10.1109/PDGC56933.2022.10053140.
24. S. S. Chawathe, 2022, "Optical Features for Automated Determination of Agricultural Product Varieties", in *2022 IEEE World AI IoT Congress (AIIoT)*, pp. 635–641, IEEExplore. doi: 10.1109/AIIoT54504.2022.9817320.
25. M. Deshmukh, A. Jaiswar, O. Joshi, R. Shedge, 2022, "Farming Assistance for Soil Fertility Improvement and Crop Prediction using XGBoost", *ITM Web of Conferences*, vol. 44, p. 03022. doi: 10.1051/itmconf/20224403022.
26. K. P. Parmar, T. Bhatt, 2022, "Crop Yield Prediction based on Feature Selection and Machine Learners: A Review". doi: 10.1109/icais53314.2022.9742891.
27. S. S. Chawathe, 2022, "Automated Determination of Mushroom Edibility Using an Augmented Dataset", in *2022 IEEE World AI IoT Congress (AIIoT)*, pp. 617–623, IEEExplore. doi: 10.1109/AIIoT54504.2022.9817321.
28. S. Dayalini, M. Sathana, P. R. N. Navodya, R. W. A. I. M. N. Weerakkodi, Anuradha Jayakody, Narmada Gamage, 2021, "Agro-Mate: A Virtual Assister to Maximize Crop Yield in Agriculture Sector", in *2021 3rd International Conference on Advancements in Computing (ICAC)*, pp. 1–6, IEEExplore. doi: 10.1109/TENCON54134.2021.9707199.
29. S. Dayalini, M. Sathana, P. R. N. Navodya, R. W. A. I. M. N. Weerakkodi, A. Jayakody, N. Gamage, 2021, "Agro-Mate: A Virtual Assister to Maximize Crop Yield in Agriculture Sector", in *TENCON 2021 – 2021 IEEE Region 10 Conference (TENCON)*, pp. 387–392, IEEExplore. doi: 10.1109/ICAC54203.2021.9671140.
30. F. Raimundo, A. Glória, P. Sebastião, 2021, "Prediction of Weather Forecast for Smart Agriculture Supported by Machine Learning," in *2021 IEEE World AI IoT Congress (AIIoT)*, Seattle, WA, USA, 2021, pp. 0160–0164, IEEExplore. doi: 10.1109/AIIoT52608.2021.9454184.
31. D. Gosai, C. Raval, R. Nayak, H. Jayswal, A. Patel, 2021, "Crop Recommendation System using Machine Learning", *International Journal of Scientific Research in Computer Science, Engineering and*

Information Technology, p. 558, TechnoScience Academy. doi: 10.32628/CSEIT2173129.

32. M. Champaneri, D. Chachpara, C. Chandvidkar, M. Rathod, 2020, "Crop yield prediction using machine learning", *International Journal of Science Research (IJSR)*, vol. 9, p. 2. doi: 10.1007/978-981-99-3010-4_38.
33. R. Ganesh Babu, C. Chellaswamy, T.S. Geetha, Daniel Raj T., K. Venkatachalam, M.A. Mulla, 2020, "Soil Test Based Smart Agriculture Management System", in *2020 7th International Conference on Smart Structures and Systems (ICSSS)*, pp. 1–6, IEEExplore. doi: 10.1177/0361198120983008.
34. C. V. Suresh Babu, 2022, *Artificial Intelligence and Expert Systems*, Anniyappa Publication
35. C. V. Suresh Babu, 2023, *IoT and its Applications*, Anniyappa Publication.

Chapter 4

Artificial Intelligence Based 3D Mapping for Soil

Kavita Patel, Priti Kumari, Vandana Dubey, Nishu Singh, Sachin Kumar, and Vijendra Pratap Singh

4.1 INTRODUCTION

A commonly known practice, agriculture or farming, involves the cultivation of crops and the rearing of livestock. This essential practice significantly contributes to the Gross Domestic Product (GDP) of a nation and serves as a major source of food and raw materials. The increasing utilization of the Internet of Things (IoT) over the last few years has had a major role in transforming traditional technology in homes, farms, and workplaces. The IoT enables not only human-to-human interaction but also facilitates human-to-device and device-to-device communication. One notable application of the IoT is in the domain of smart farming, which leverages emerging technologies to enhance agricultural productivity and product quality. This setup integrates components like soil analysis, temperature and humidity tracking, and data retention. The adoption of smart farming practices has proven beneficial for both tiny and large-scale farmers, providing them with devices to access cutting-edge methodologies that optimize crop yield and quality while reducing production costs. The utilization of raw resources such as cotton and jute in various industries plays a crucial role in the creation of everyday items [1]. Agriculture not only supports food production but also provides the necessary materials for the manufacturing of goods. Traditional agricultural methods were employed to cultivate crops, which predominantly remain the conventional approach worldwide. These time-consuming methods, recommended by experienced farmers, require extensive manual labor due to their lack of precision [2].

In contrast, precision agriculture embraces digital technology, including robotics, electronics, sensors, and automation, aiming to optimize operations, enhance productivity, and facilitate informed decision-making. Precision agriculture effectively increases crop yield and adoption rates vary depending on the value and geographical location of the farms [3]. Previously, individuals managing household chores and/or occupied with professional liabilities had limited awareness of the environmental situations outside of their homes in the absence of the IoT. They lacked knowledge about the external

DOI: 10.1201/9781003676102-4

temperature (whether it was high, low, or normal), rainfall, or humidity levels. In such scenarios, IoT devices prove extremely useful by providing alerts regarding extreme temperatures. High-value enterprise farms tend to have a higher adoption rate compared to low-value enterprise farms.

However, precision agriculture adoption is relatively lower in mountainous regions, primarily due to the significant investments required. Therefore, reducing the cost of machinery is crucial to enable all types of farms to benefit from precision agriculture. Modern technologies like Data Mining, the IoT, Data Science, and Artificial Intelligence play a vital role in advancing precision agriculture. The IoT refers to a network of interconnected computational objects, such as smart devices and sensors, that can store and exchange data with each other [4]. The application of smart farming equips farmers with real-time information such as moisture of soil, humidity levels, temperature, and water availability, guiding them as they adopt traditional agricultural practices that enhance crop yield, quality, and overall production.

Wireless Sensor Networks (WSNs) are employed in agronomic applications to remotely monitor environmental and soil variables, enabling the prediction of crop health and forecasting irrigation schedules. WSNs collect data not only on pressure, humidity, and temperature but also on soil moisture, salinity, and conductivity, facilitating efficient agricultural practices. The estimation of rainfall in advance is crucial for effective crop management. Additionally, soil plays a vital role in agriculture, affecting factors such as anchoring, moisture and nutrient storage, mineral reserves, and root growth [5]. Soil testing is an essential step in soil preparation, helping determine existing nutrient levels and the appropriate amount of nutrients required based on soil fertility and crop needs. Soil testing categorizes important soil factors, including phosphorus, potassium, nitrogen, organic carbon, boron, and soil pH [6]. Smart farming envisions a promising future, introducing farming technologies that aim to reduce costs, enhance farming efficiency, and deliver high-quality products. Additional advantages of intelligent agriculture encompass lessening environmental stress, enhancing the well-being of farmers and laborers, reducing the release of chemicals into rivers, and managing water consumption.

Irrigation, a critical aspect of agriculture, contributes to soil and water conservation. Employing complex data analysis considering water availability, soil conditions, climate, and crop characteristics, irrigation systems can be optimized for performance and uniformity. Deep learning has emerged as a powerful solution in this regard, providing exceptional results when tackling the complexity of irrigation systems. The cultivation of crops involves various steps, and this work emphasizes the use of Deep Learning and Machine Learning techniques to address numerous challenges in the field [7]. Overall, the integration of IoT and smart farming offers numerous advantages, contributing to sustainable and efficient agricultural practices while fostering environmental stewardship and resource management.

The subsequent sections of this chapter are organized as follows: Methodology is described in Section II, while Section III presents the results and experimental exploration of the proposed technique. Future directions and concluding remarks are highlighted in Section IV.

4.2 RELATED WORK

The system for monitoring and control is made up of an Android and multiple sensors, as shown in this article by Sridharani et al [1]. The database system used hardware for its input and provided an analysis of the data to advise the farmer. The system suggested what organic crop might be best in this kind of soil, what irrigation techniques, optimal farming techniques, and so forth. This program uses a different Android app that farmers might not be aware of. Sangeetha et al [2] mention guiding farmers by making some modifications to raise the productivity or the whole crop. For farmers, this system updates soil-related information. There are no reliable internet connections in the villages because this project depends on an internet connection to function.

Sujatha et al [3] project, describe a technology-based system for tracking ground characteristics, such as soil moisture and temperature, as well as environmental factors in order to estimate the watering requirements for a particular field. The internet can be used to access parameters like weather forecast information. Farmers who use this kind of irrigation technology can accomplish their goals while using the least amount of water possible. This system does not fully utilize all types of sensors. Mutyalamma et al mention a system that uses the results to safeguard plants from fire, allowing us to exercise caution beforehand. This gradually contributes to less crop devastation. Additionally, the system makes a big dent in the fight against global warming. A wireless sensor network was used by this system. A lot of technology is employed, including cameras, to capture remote circumstances, which may not be achievable in rural areas.

Velmurugan et al demonstrate how to monitor the temperature and pressure of a crop while it is being transported from one location to another. Updates from retail and wholesale stores regarding the farmer's crop consumption can also be obtained using it. Unwanted hardware effects could have a negative impact on the outcomes. A separate GSM module is attached to the Arduino by Ma et al [6] to offer an SMS facility and report the weather to the farmer every 10 seconds. Additionally, this device tracks the PH values of the soil, plants, soil moisture, and temperature. This depends on a constant GPS signal. Remote locations might not be able to do this.

Swarmi et al [7] illustrate difficulties with network-based electronic device control. The data offered by the sensors can be monitored and analyzed from any part of the world via the internet. To obtain updates on the soil and the plants, sensors were installed on every plant in that field, as shown in Ritika et al [8]. This device does not examine all the parameters; it merely measures the moisture content of the soil. Since it needs to be placed for each plant on the field, its application on a broad scale is expensive.

Durai and Shamili [9] highlight the strategies that must be used to stop the overuse of water resources. The major goal is to focus on locations with water shortages to determine what crops they can plant and how much water is needed to grow those crops, and so forth. Farmers who use this technology must be knowledgeable about microcontrollers, which some farmers may not have. The technology discussed by Franz et al [10] assists farmers in practising smart farming by enhancing agricultural productivity and resource conservation, such as fertilizers, water and pesticides. Moreover, it effectively reduces the wastage of crops while being cost-efficient. It achieves accurate results using a minimal number of sensors, detecting all necessary field conditions.

Mythili et al [11] delve into the challenge that comes with incorporating Internet of Things (IoT) technologies into conventional farming methods. Some of these challenges include the possibility of sensor failure, leading to inaccurate data forecasts, and potential errors in the farmer's input to the field, which can impact crop yield. It is specified by Lalbihari [12], which crops are ideal for a given type of soil type and which crops produce a higher growth when they are grown. Soil characteristics include things like moisture, acidity pH, and many more. Additionally, this aids the farmer in lowering input costs and minimizing environmental damage that could harm the crop or diminish production. The output that the system produces does not distinguish between different colors. This could cause difficulties for farmers who lack the expertise needed to understand the differences and provide the right input for a crop to achieve a high yield.

A paper by Gayathri et al [13] employs sensors placed in the field to detect the weather and irrigate the plants accordingly, thus automating the farmer's labor. The benefit or added feature of this system is that it monitors each step of crop cultivation or harvesting uniquely. As in Kumar et al [14], the system predicts soil moisture content and efficiently waters plants, leading to reduced water resource usage, as water is a crucial component in the farming process. It proves to be user-friendly in rural places and requires minimal maintenance within limited-scale applications. Managing the vast amount of information in such cases also poses challenges. Doshi et al [15] highlight the challenges arising from internet connections in this context. Inaccurate field conditions and soil characteristic information can arise due to communication disruptions between sensors and other electronic devices. Their paper discusses these challenges extensively. Nevertheless, it is important to note that while the challenges are addressed in the paper, a real-time application for determining soil characteristics has not been developed yet.

The IoT is the newest revolution of the internet age, and it streams live data on the status of the entire planet, including information on temperature, rainstorms, moisture, cataracts, earthquakes, and other occurrences that could threaten the viability of human life. Anitha et al [16] provide an inexpensive rainfall observation device that can obtain the current rainfall conditions at any location using a simple database operating system and display this on an OLED screen. The main objective of this project is to enable users to probe current data from any station and assess rainfall conditions from any location.

The system proposed Ayaz et al [17] is a state-of-the-art solution for rainfall monitoring that makes IoT enabled real-time data readily available to access across a range of devices. This entails using a variety of monitors to keep an eye on things like temperature, wetness, humidity, wind speed, light intensity, UV radiation, and even airborne levels of carbon monoxide as they relate to rainfall and climatic changes. Data gathered after the analysis from all around the globe about the way to make choices about the conditions is greatly facilitated by the Internet of Effects. Due to its similarly unique character, security is a significant challenge to protect it against colorful sorts of attacks, as noted by Palanivel and Surianarayanan [18]. There have been several security improvement approaches proposed in the past to deal with the issue of numerous attacks, but since new attacks are discovered daily, there isn't an equal method that provides total security. The general text is vivid and icy-cold.

The hygrometer alerted farmers' cell phones through GSM when the humidity of the soil went lower than a preset threshold. Villanueva and Salenga [19] discuss an IoT-connected remote-controlled truck that was developed to keep track of soil moisture, temperature, and other factors so that irrigation decisions could be made. The fundamental principles of IoT-based smart husbandry are detailed in the papers by Rakhee et al and Nevavuori et al [20, 21]. Additionally, they delve into the binding of IoT-driven husbandry techniques to pertinent methodologies like big data storage, edge computing, and analytics. The significance of ensuring security in IoT husbandry is also underscored. Furthermore, the paper includes a compilation of smartphone-based and sensor-based operations tailored for different aspects of ranch operations.

A few of the current models need the farmers to upload/install an application to look into the environment, and only a small number of others call for a continuous internet connection. This could not be accurate because many farmers don't have smartphones and rural areas might have frequent problems that farmers would be able to address. A few other systems, in contrast, present the results as text, which some farmers might be unable to comprehend. These are the main issues with the current models.

4.3 PROPOSED METHODS

The Smart Farm System comprises hardware and software elements. The hardware component encompasses embedded systems, while the software component utilizes Arduino technology. The Arduino software processes data from various sensors that are interconnected through hardware. These sensors, which include specialized ones for temperature, soil humidity, and soil moisture, transmit data to an Arduino. The information collected by these detectors is transmitted to the Raspberry Pi's Arduino U-N-O microcontroller. The collected information could be displayed on an Arduino display. The proposed work consists of four components. It gauges the soil temperature and soil humidity of the surrounding air and then transmits this data to the computer model. Additionally, it can oversee plant or soil moisture levels, as it can effectively sense the moisture close to the sensor.

The arrangement of modules are depicted which are adopted in Figure 4.1. Our initial assessment of various sensors involved utilizing dht11 and YI-69 sensors to capture temperature, humidity, and moisture data. The dht11 sensor is responsible for detecting the temperature and humidity of the surroundings, while the YI-69 sensor is employed to gauge soil moisture levels. The sensors will be independently connected to the North along with South poles of an Arduino on one side. The Raspberry Pi will be attached

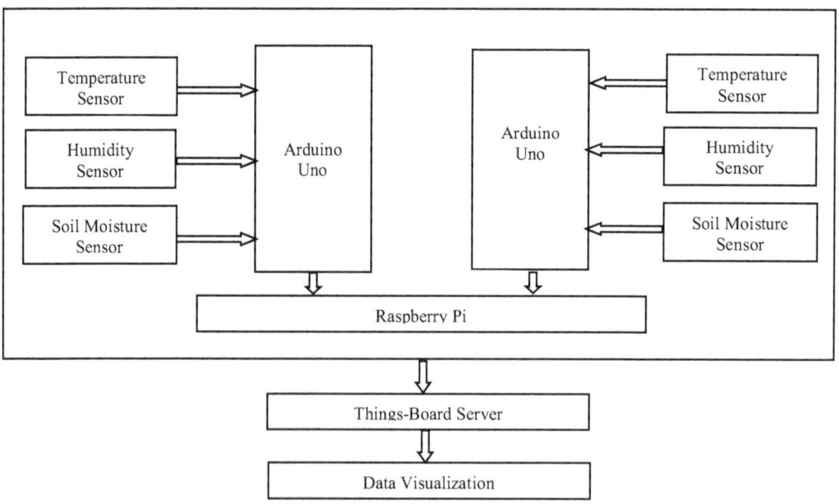

Figure 4.1 Proposed Method Architecture.

to the Arduino's opposite side, making it easier to receive information from the one and the other Arduinos and transmit it to the Things-Board Server later. The information will be transformed into a python file by the Things-Board-Server, making the data-visualization process easier. After conversion, the data can be presented in graphical form for farmers to easily interpret (Figure 4.1).

4.3.1 Temperature Sensor

The initial module employs a dht11 sensor to detect the prevailing temperature. Subsequently, the soil-temperature information will be transmitted to the MQTT broker, which in this case is the Raspberry-Pi. dht11 working: This temperature sensor generates a precise digital temperature reading and operates through a thermistor to assess the ambient temperature. It boasts a swift response time and exceptional accuracy.

4.3.2 Humidity Sensor

The second module employs a dht11 to measure the current humidity level. The Raspberry Pi, which serves as the MQTT broker, will receive temperature input. dht11 working: It's a calibrated digital humidity output produced by a humidity sensor. The surrounding environment is measured with this humidity detector/sensor, with achieving a rapid response time and maintaining high quality standards being notable characteristics.

4.3.3 Soil Moisture Sensor

In the THIRD module, an Arduino is employed as a microcontroller to acquire soil moisture data from a soil-moisture sensor and subsequently transmit it to an MQTT broker. Yl-69 working: It is a soil moisture detector that gauges how much water is present in the ground. It can determine whether the soil is too moist or too dry. The soil's resistance is read after the two probes have passed current through it to ascertain the soil's moisture content.

4.3.4 Data Visualization

We are visualizing data using Things-Board, a fantastic open source IoT platform for real-time data-visualization. Data-Subscriber. py working: The role of this code is to access information from the agent across multiple places and record it in diverse text-files. Within this context, Things-Board north.py and Things-board south.py represent two distinct Python modules. Both of these scripts retrieve info from the text-files and proceed to load it to the Things-Board server using a Python file.

4.3.5 Process of System Flow

Sensors for soil-temperature, moisture content, and soil-humidity are strategically kept at the South and North poles of the farm to gauge these variables. Subsequently, all the acquired soil moisture, soil- temperature, and humidity values are forwarded to the software on an Arduino device board, where the data is processed based on the outcomes. The monitoring is carried out through the Things-Board Server, facilitating the visualization of data. This visualization aids the farmer in understanding the environmental conditions and making informed decisions (Figure 4.2).

Different places have varying levels of temperature, wetness, and humidity. Conclusions that are drawn purely based on readings from a specific location on the farm may be incorrect or inappropriate. Therefore, in order to obtain a more accurate value, we install three sensors at opposite ends of the farm. Both the North and South poles are equipped with these sensors and Arduino boards. We learn more about the general characteristics by averaging the values at these two poles.

The connections and interactions of the server model are comprehensively elaborated upon in Figure 4.2. Information is gathered such as soil-temperature, soil-moisture, and soil-humidity from distinct locations within the field, namely South and North points, after that, they present this information on the server of Things-Board. In Figure 4.3, we track the results at the field's North and South poles to get the most precise results. Having access to this technology and integrating it into their farms is essential for farmers. Both the system and the sensors operate efficiently. It is simple

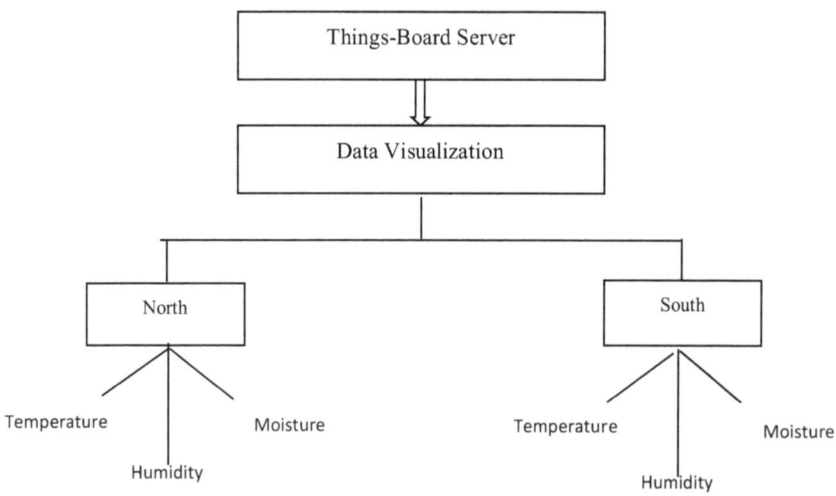

Figure 4.2 System Flow Process.

to install and maintain. While it requires minor upkeep, the maintenance costs are minimal compared to the benefits it offers. The sensors capture data from the surrounding environment and transmit this information to the Arduino board.

To obtain reliable sensor findings, the Arduinos are put at the field's poles and average the results. The Arduino processes the obtained data and displays it on the Arduino IDE. In order for the farmer to see them, the data or values are also sent to the Things-Board sensor. The inclusion of the Things-Board sensor within the system serves as a sufficient solution in this context, eliminating the necessity for a smartphone or conventional phone for the farmer. To enhance the farmer's understanding of the outcomes and facilitate informed decision-making, the results are visualized through graphs and diagrams, employing distinct colors for clear differentiation.

To ensure the accuracy of the results acquired, we selected two places on the ground to calculate the soil properties in the proposed system: one at the South pole and another at the North pole. If the field has a slope, using just one site to measure the features could result in different numbers for the entire field; as a result, we used two points to forecast the output on average. Another point of comparison is that we avoid the utilization of separate software for displaying outputs, which could potentially confuse farmers regarding its operation. This system, on the other hand, benefits farmers by providing outputs presented in graphs with color differentiation. This approach eliminates any ambiguity for farmers when interpreting the displayed results.

4.4 RESULTS AND DISCUSSIONS

An IoT based 'Smart Farming System' for real-time temperature monitoring, checking soil-humidity, and soil-moisture has been shown using Arduino and sensors. The proposed smart farming system based on the IoT, as outlined in this research, holds the potential to assist farmers in enhancing agricultural yields and ensuring effective crop management. This system serves as a reliable resource, continuously offering real-time and precise environmental temperature and soil moisture data, thereby providing valuable support to farmers.

Figure 4.3 uses a bar graph to show the properties of the soil that was gathered by the Arduinos placed at both the North and South poles. The North pole is represented by the color grey in the graph, whereas the South pole is represented by the color black. Soil moisture, humidity, and temperature contents at both poles are among the soil factors that were measured.

The temperature variations in Celsius over a specific time frame are visualized in the graph presented in Figure 4.4. The black plot represents the minimum, average and maximum temperatures recorded at both the

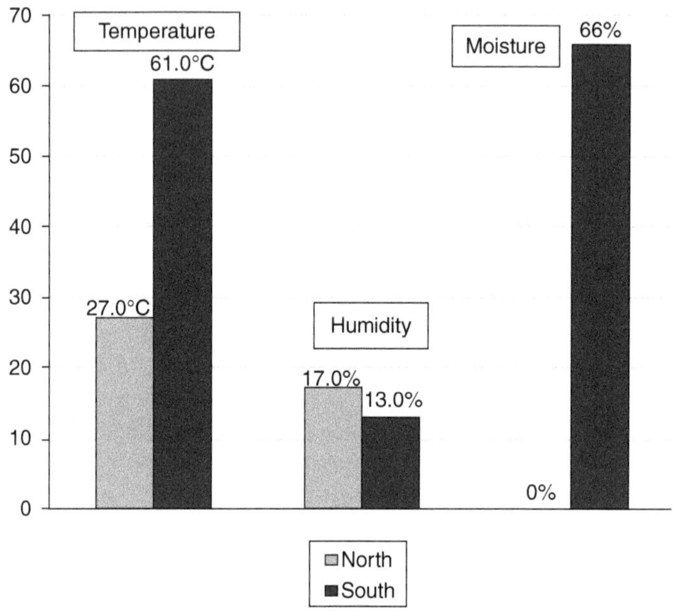

Figure 4.3 North and South Poles Soil Characteristics.

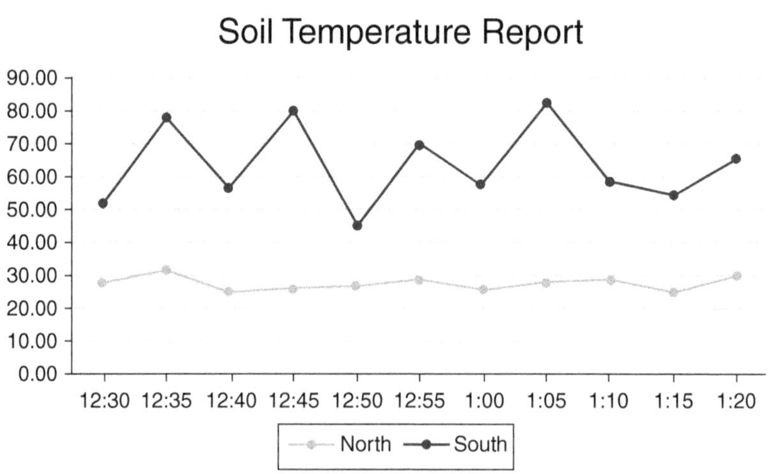

Figure 4.4 Soil Temperature Report.

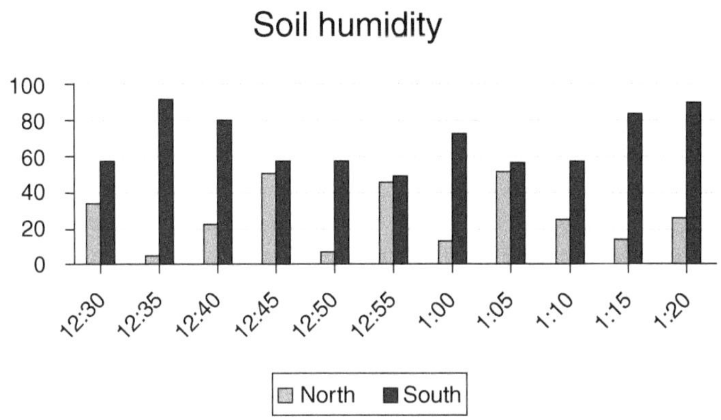

Figure 4.5 Soil Humidity Report.

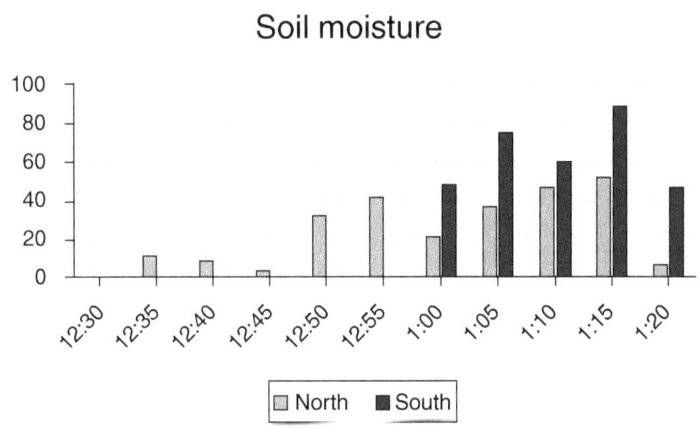

Figure 4.6 Soil Moisture Report.

North and South poles during this period. Meanwhile, the grey hue on the graph illustrates the temperature fluctuations specifically at the North pole.

Humidity is the next crucial characteristic to consider, as shown by the pictorial graph in Figure 4.5, which shows the percentage changes in soil-humidity over a given time period at both the South and North poles.

The bar-graph in grey displays variation in humidity at the North-side pole, while the given graph in black displays the fluctuation in humidity at the South-side pole. The average of soil-humidity at the North and South poles is likewise affected by this.

Crop growth greatly depends on the moisture level of the soil. The graph represented in Figure 4.6, which shows the percentage of soil-moisture content at the North and South poles over a set period of time, emphasizes this. The South pole is symbolized by the color black, whereas the North pole is indicated by the color grey. An interesting finding from the graph is that the moisture content is higher in the South pole, which is explained by the flow of water from higher to lower elevations within the designated geographical area.

The absence of crop screening leads to a dearth of visual data, consequently failing to identify issues related to plants. The emphasis is primarily on field circumstances rather than plant conditions. Since it is constructed on an existing server infrastructure, communication isn't extended through Wi-Fi or cellular networks. Data is gathered through sensors, providing exclusively numerical digital readings.

4.5 CONCLUSION

This research presents a comprehensive IoT-based SMART FARMING SYSTEM that leverages Arduino and sensors for real-time monitoring of soil-temperature, soil-humidity and soil-moisture. The system's implementation has demonstrated exceptional accuracy and efficiency in retrieving critical environmental data, with a focus on the North and South poles. The significance of this smart farming system is underscored by its potential to revolutionize agricultural practices. By providing farmers with real-time and precise information about temperature, humidity, and soil moisture, the system equips them with the tools needed to enhance crop management and increase agricultural yields.

The research highlights the role of moisture content, especially in influencing crop growth, and showcases how the data collected from both poles contributes to a better understanding of soil conditions. Overall, this IoT-based smart farming system presents a promising approach for modernizing agriculture through technology. Its capacity to provide continuous and accurate environmental data can potentially drive informed decision-making, leading to improved crop yields and sustainable farming practices. As technology continues to advance, integrating IoT solutions like this into agriculture could pave the way for a more efficient and productive future for farmers around the world.

4.6 FUTURE SCOPE

The acquired data has the potential to create a sizable dataset that is useful for developing a predictive learning model that aims to foresee future changes in field features. The collecting of picture data might be made easier by expanding the array of sensors and introducing screening systems throughout the field to improve accuracy in recognizing plant-related diseases and problems. Direct data transfer to the farmer's mobile device, which need not be a smartphone in particular, could be made possible by the integration of a Wi-Fi module. Incorporated within this system is an intelligent module, which will assess the frequency of watering for specific crops in accordance with prevailing weather conditions. It will then calculate the optimal water volume required to safeguard the crops from harm. By employing a combination of camera motion sensing and advanced machine learning techniques, the system can even detect unfamiliar elements, such as potential crop threats from animals. In such instances, it can capture an image and dispatch a warning message to the farmer as a precautionary measure.

REFERENCES

1. J. Sridharani, S. Chowdary and K. Nikhil, "Smart Farming: The IoT based Future Agriculture," in *2022 4th International Conference on Smart Systems and Inventive Technology (ICSSIT), Tirunelveli, India*, 2022, pp. 150–155, doi: 10.1109/ICSSIT53264.2022.9716331.
2. K. Sangeetha, C. Narmada, D. Moneeshram, R. Karishma and V. Kishore Karthi, "Smart Farming Using IoT", *Special Issue of Second International Conference on Advancements in Research and Development (ICARD 2021)*, vol. 3, no. 03S, pp. 63–67, Mar. 2021.
3. M. Sujatha and C. D. Jaidhar, "Classification of Soil Fertility using Machine Learning-based Classifier," in *2021 2nd International Conference on Secure Cyber Computing and Communications (ICSCCC), Jalandhar, India*, 2021, pp. 138–143, IEEE. doi: 10.1109/ICSCCC51823.2021.9478169.
4. A. V. Mutyalamma, G. Yoshitha, A. Dakshyani and B. V. Padmavathi, "Smart Agriculture to Measure Humidity, Temperature, Moisture, Ph. of the Soil using IoT", *International Journal of Engineering and Advanced Technology (IJEAT)*, vol. 9, no. 5, 394–398, June 2020, ISSN: 2249 – 8958.
5. S. Velmurugan, V. Balaji, T. Manoj Bharathi and K. Saravanan, "An IOT Based Smart Irrigation Using Soil Moisture and Weather Prediction", *International Journal of Engineering and Advanced Technology (IJEAT)*, vol. 9, no. 2, 1-4, Apr. 2022, E-ISSN 2348-1269, P- ISSN 2349-5138.
6. C. Ma, X. Li and M. F. McCabe, "Retrieval of High Resolution Soil Moisture through Combination of Sentinel-1 and Sentinel-2 Data", *Remote Sensors*, vol. 12, no. 14, p. 2303, Jul. 2020.
7. C. M. Swaraj and K. M. Sowmyashree, ""IOT Based Smart Agriculture Monitoring and Irrigation System", *International Journal of Engineering Research & Technology* (IJERT), vol. 9, no. 4, 245–249, Apr. 2020, ISSN: 2249 – 8958 (Online).
8. S. Ritika, S. Vandana, J. Vishal and R. Sumit, "A Research Paper On Smart Agriculture Using IoT", *International Research Journal of Engineering and Technology (IRJET)*, vol. 7, no. 7, 2708–2710, Jul. 2020, e-ISSN: 2395-0056, p-ISSN: 2395-0072.
9. S. K. S. Durai and M. D. Shamili, "Smart Farming using Machine Learning and Deep Learning Techniques", *Decision Analytics Journal*, vol. 3, p. 100041, 2022, ISSN 2772-6622, doi: 10.1016/j.dajour.2022.100041.

10. T. E. Franz, A. Wahbi, J. Zhang, M. Vreugdenhil, L. Heng, G. Dercon, et al., "Practical Data Products from Cosmic-Ray Neutron Sensing for Hydrological Applications", *Frontiers in Water*, vol. 2, p. 9, 2020.
11. R. Mythili, M. Kumari, A. Tripathi and N. Pal, "IoT Based Smart Farm Monitoring System", *International Journal of Recent Technology and Engineering* (IJRTE), vol. 8, no. 4, pp. 5490–5494, ISSN: 2277-3878 (Online).
12. Lalbihari Barik, "IoT based Temperature and Humidity Controlling using Arduino and Raspberry Pi" *International Journal of Advanced Computer Science and Applications (IJACSA)*, vol. 10, no. 9, 494–502, 2019. doi: 10.14569/IJACSA.2019.0100966.
13. M. Gayathri, D. Shunmugam and A. Ishwariya, "Smart Irrigation System using IoT", *International Journal of Innovative Research in Engineering & Multidisciplinary Physical Sciences*, vol. 08, no. 10, 149–154, Oct. 2021.
14. A. Kumar, S. Sarkar and C. Pradhan, "Recommendation System for Crop Identification and Pest Control Technique in Agriculture," in *2019 International Conference on Communication and Signal Processing (ICCSP)*, Chennai, India, 2019, pp. 0185–0189, doi: 10.1109/ICCSP.2019.8698099..
15. J. Doshi, T. Patel and S. K. Bharti, "Smart Farming using IoT, A Solution for Optimally Monitoring Farming Conditions", *Procedia Computer Science*, vol. 160, pp. 746–751, 2019, ISSN 1877-0509, doi: 10.1016/j.procs.2019.11.016.
16. A. Anitha, N. Sampath and A. Jerlin, "Smart Irrigation System using Internet of Things", *International Conference on Emerging Trends in Information Technology and Engineering (IC-ETITE)*, Publisher: IEEE, Xplore Place: Vellore, India, 978-1-7281-4142-8/20/$31.00 ©2020 IEEE 10.1109/ic-ETITE47903.2020.271.
17. M. Ayaz, M. Ammad-Uddin, Z. Sharif, A. Mansour and E. -H. M. Aggoune, "Internet-of-Things (IoT)-Based Smart Agriculture: Toward Making the Fields Talk," *IEEE Access*, vol. 7, pp. 129551–129583, 2019, doi: 10.1109/ACCESS.2019.2932609.
18. K. Palanivel and C. Surianarayanan, "An Approach for Prediction of Crop Yield Using Machine Learning and Big Data Techniques", *International Journal of Computer Engineering and Technology*, vol. 10, no. 3, pp. 110–118, 2019.
19. B. M. Villanueva and M. L. M. Salenga," Bitter Melon Crop Yield Prediction using Machine Learning Algorithm", *International Journal of Advances Computer Science Applications*, vol. 9, pp. 1–6, 2018.
20. K. Rakhee, R. V. Ohri, Saraswathi and L. J. Vinita, "Performance Analysis of Wireless Body Area Sensor Analytics using Clustering Technique," in *2019 International Conference on Communication and Signal Processing (ICCSP)*, 2019, pp. 0278–0281, doi: 10.1109/ICCSP.2019.8697963.
21. P. Nevavuori, N. Narra and T Lipping, "Crop yield prediction with deep convolutional neural networks", *Computers and Electronics in Agriculture*, vol. 163, p. 104859, 2019.

Chapter 5

Machine Learning for Data-Driven Agriculture

Y. Jahnavi, P. Nagendra Kumar, B. Yamini, K. SandeepuKumar, P. Anusha, and M. Hymavathi

5.1 INTRODUCTION

Data-driven agriculture, also known as precision agriculture or smart farming, is revolutionizing the way farming practices are conducted. Traditional farming methods often relied on subjective decision-making based on experience and intuition. However, with the advancements in sensing technologies and the availability of large-scale agricultural data, machine learning has emerged as a powerful tool for making informed and data-driven decisions in agriculture [1].

Machine learning involves the development of algorithms and models that can learn patterns and make predictions or decisions based on data. In the context of agriculture, machine learning algorithms can analyze vast amounts of agricultural data, including soil properties, weather conditions, crop characteristics, and yield information, to extract valuable insights and optimize various aspects of farming operations.

The availability of agricultural data has significantly increased in recent years. Technologies such as remote sensing, which uses satellite imagery and aerial photography, provide detailed information about crop health, growth patterns, and soil moisture levels. Internet of Things (IoT) devices deployed in the field can collect real-time data on temperature, humidity, rainfall, and other environmental variables. Unmanned Aerial Vehicles (UAVs) equipped with sensors can capture high-resolution images of crops and monitor their condition [2].

By leveraging this wealth of data, machine learning algorithms can help farmers address key challenges in agriculture, such as optimizing irrigation and fertilizer application, predicting crop yield, detecting and managing pests and diseases, and improving overall resource management. Machine learning models can learn from historical data and patterns to make accurate predictions and recommendations for future farming practices.

Moreover, the integration of machine learning with other agricultural technologies, such as precision agriculture systems and farm management

DOI: 10.1201/9781003676102-5

software, allows for the seamless flow of data and enables comprehensive decision support systems for farmers. These systems can provide real-time recommendations and insights, empowering farmers to make timely and informed decisions to maximize crop productivity while minimizing costs and environmental impact [3–6].

However, the implementation of machine learning in agriculture is not without challenges. Agricultural data can be diverse, heterogeneous, and prone to noise and missing values. Data collection and integration can be complex, requiring careful consideration of data quality and reliability. Furthermore, machine learning models need to be interpretable and transparent to gain the trust of farmers and stakeholders who rely on their recommendations.

In this chapter, we will explore various aspects of machine learning for data-driven agriculture. We will discuss different machine-learning techniques applicable to agriculture, including supervised learning, unsupervised learning, and reinforcement learning. We will delve into data pre-processing and feature engineering techniques specific to agricultural datasets. Additionally, we will explore case studies and real-world applications of machine learning in agriculture.

We first discuss the different types of agricultural data and its sources, highlighting the challenges associated with data collection, integration, and quality assurance. We then delve into various machine learning techniques commonly employed in agriculture, including supervised learning, unsupervised learning, and reinforcement learning.

Supervised learning algorithms, such as random forests, support vector machines, and neural networks, have been widely used for tasks such as crop yield prediction, disease detection, and weed classification. Unsupervised learning methods, such as clustering and dimensionality reduction, enable data exploration, anomaly detection, and pattern recognition in agricultural datasets. Reinforcement learning algorithms provide a framework for optimizing sequential decision-making processes, such as irrigation scheduling and pest control strategies [7, 8].

While the benefits of machine learning in agriculture are evident, challenges remain. Data privacy, data ownership, interpretability of models, and limited access to technology in rural areas are among the challenges that must be addressed for the widespread adoption of data-driven agriculture. Additionally, ongoing research and development are necessary to address the specific needs and nuances of the agricultural domain.

Furthermore, we discuss the importance of data pre-processing and feature engineering in preparing agricultural data for machine learning models. Techniques such as data cleaning, normalization, feature selection, and extraction play a vital role in improving model accuracy and generalization [9–17]. We also explore the integration of machine learning models with other agricultural technologies, such as precision agriculture systems and farm management software, to create comprehensive data-driven farming solutions.

Finally, we highlight the benefits and challenges of implementing machine learning in agriculture. The benefits include improved crop yield, resource optimization, early pest detection, and reduced environmental impact. However, challenges such as data privacy, data ownership, interpretability of models, and limited access to technology in rural areas need to be addressed for the widespread adoption of data-driven agriculture.

By leveraging the power of machine learning, data-driven agriculture has the potential to transform farming practices and contribute to sustainable and efficient food production. By optimizing resource allocation, improving crop management, and mitigating risks, machine learning can help farmers make better decisions, increase productivity, and contribute to global food security in an environmentally conscious manner [18].

5.2 LITERATURE REVIEW

Several papers illustrate the key features of machine learning algorithms to improve crop yield. The authors represent an overview of machine learning applications in agriculture. Various topics are covered, including crop yield prediction, disease detection, weed classification, and soil mapping. The authors discuss different machine learning algorithms and techniques employed in agricultural applications and highlight the challenges and future directions of the field [19, 20, 21].

The authors provide insights into the potential benefits and limitations of machine learning in agriculture. This review explores the utilization of machine learning in agriculture with a focus on crop production and management. It discusses the application of machine learning techniques in predicting crop yields, optimizing irrigation and fertilization, and detecting crop diseases and pests [22–26].

The authors discuss the challenges associated with disease detection in agricultural systems and highlight the potential of machine learning in improving crop health monitoring and management. This chapter focuses on the application of machine learning techniques for crop disease detection and diagnosis. It provides an overview of different machine learning algorithms used to identify and classify plant diseases based on image analysis.

The authors discuss the challenges and future directions of deep learning in remote sensing for agriculture [27–31]. This tutorial chapter focuses on applying deep learning techniques, a subset of machine learning, for remote sensing data analysis in agriculture. It provides an in-depth explanation of deep learning architectures, such as Convolutional Neural Networks (CNNs) and Recurrent Neural Networks (RNNs), and their applications in crop classification, yield prediction, and land cover mapping [32, 33, 34].

These literature review papers provide valuable insights into the application of machine learning in data-driven agriculture. They cover a wide range of topics, including crop yield prediction, disease detection, pest management, and remote sensing analysis. By examining the different algorithms, techniques, and challenges in the field, these papers contribute to understanding and advancing machine learning in agriculture [35, 36].

5.3 MODEL BUILDING AND EXPERIMENTAL ANALYSIS

The dataset with the attributes Year, Location, Soil Type, Max_ Temperature, Min_ Temperature, Rainfall, Humidity, and Yield Kg/ha were considered for the experimentation. The considered dataset contains 8077 instances and 8 attributes. Among the 8 attributes, Year, Max_ Temperature, Min_ Temperature, Rainfall, Humidity, and Yield Kg/ha are numeric attributes, and Location and Soil Type are nominal attributes. A sample dataset has been represented in Table 5.1.

Year attribute can be dropped as it is not useful for the prediction process and represented in Table 5.2.

Table 5.1 Sample Dataset

S. No	*Year*	*Location*	*Soil Type*	*Max_ Temperature*	*Min_ Temperature*	*Rainfall*	*Humidity*	*Yield Kg/ha*
1	2018	Alemtena	Light Soil	29.0	12.0	543.8	46.0	2830.0
2	2018	Denkaka	Black Soil	28.5	13.0	543.8	45.0	2485.0
3	2019	Chefe Donsa	Black Soil	30.0	12.5	811.2	58.0	240.0
4	2013	Minjar	Light Soil	26.0	8.0	487.6	67.0	1458.0
5	2019	Chefe Donsa	Black Soil	30.0	13.0	811.2	51.0	470.0
6	2019	DZ-BS	Black Soil	30.0	13.5	811.2	45.0	720.0
7	2017	Alemtena	Light Soil	26.5	11.0	612.0	60.0	420.0
8	2019	DZ-BS	Black Soil	29.5	12.0	811.2	54.0	4020.0
9	2019	DZ-BS	Black Soil	26.0	10.2	811.2	58.0	3510.0
10	2014	Alemtena	Light Soil	24.5	15.0	559.8	46.0	1195.0
11	2019	Chefe Donsa	Black Soil	31.0	13.0	811.2	68.0	980.0
12	2018	Chefe Donsa	Black Soil	30.5	16.0	811.2	72.0	670.0
13	2017	Chefe Donsa	Black Soil	29.5	14.0	811.2	68.0	730.0
14	2016	Chefe Donsa	Black Soil	30.0	13.0	811.2	54.0	710.0
15	2019	Chefe Donsa	Black Soil	31.0	12.0	811.2	70.0	780.0

Table 5.2 Sample Dataset with the Required Attributes

S. No.	*Location*	*Soil Type*	*Max_ Temperature*	*Min_ Temperature*	*Rainfall*	*Humidity*	*Yield Kg/ha*
1	Alemtena	Light Soil	29.0	12.0	543.8	46.0	2830.0
2	Denkaka	Black Soil	28.5	13.0	543.8	45.0	2485.0
3	Chefe Donsa	Black Soil	30.0	12.5	811.2	58.0	240.0
4	Minjar	Light Soil	26.0	8.0	487.6	67.0	1458.0
5	Chefe Donsa	Black Soil	30.0	13.0	811.2	51.0	470.0
6	DZ-BS	Black Soil	30.0	13.5	811.2	45.0	720.0
7	Alemtena	Light Soil	26.5	11.0	612.0	60.0	420.0
8	DZ-BS	Black Soil	29.5	12.0	811.2	54.0	4020.0
9	DZ-BS	Black Soil	26.0	10.2	811.2	58.0	3510.0
10	Alemtena	Light Soil	24.5	15.0	559.8	46.0	1195.0
11	Chefe Donsa	Black Soil	31.0	13.0	811.2	68.0	980.0
12	Chefe Donsa	Black Soil	30.5	16.0	811.2	72.0	670.0
13	Chefe Donsa	Black Soil	29.5	14.0	811.2	68.0	730.0
14	Chefe Donsa	Black Soil	30.0	13.0	811.2	54.0	710.0
15	Chefe Donsa	Black Soil	31.0	12.0	811.2	70.0	780.0

Table 5.3 Subsample Dataset

S. No.	*Location*	*Soil Type*	*Max_ Temperature*	*Min_ Temperature*	*Rainfall*	*Humidity*	*Yield Kg/ha*
1	Chefe Donsa	Black Soil	31.0	13.0	811.2	68.0	980.0
2	Chefe Donsa	Black Soil	30.5	16.0	811.2	72.0	670.0
3	Chefe Donsa	Black Soil	29.5	14.0	811.2	68.0	730.0
4	Chefe Donsa	Black Soil	30.0	13.0	811.2	54.0	710.0
5	Chefe Donsa	Black Soil	31.0	12.0	811.2	70.0	780.0

Model building was undertaken by using various ensemble learning algorithms.

Before building and training the model, the dependent variable and independent variables need to be separated and then the dataset split into the training and testing sets. Location, Soil Type, Max_Temperature, Min_Teperature, Rainfall, and Humidity are independent features. Yield Kg/ha is a dependent feature. Subsample datasets (Figure 5.3) were considered for separating dependent (Figure 5.4) and independent (Figure 5.5) features.

Data Transformation: Data Transformation is a pre-processing technique used for changing categorical values to numeric values. In the considered dataset the dummy variables are created by changing the categorical value to a numeric value (0, 1). The transformed dataset is represented in Table 5.6.

Table 5.4 Dataset with Dependent Features

S. No.	*Yield Kg/ha*
1	980.0
2	670.0
3	730.0
4	710.0
5	780.0

Table 5.5 Dataset with Independent Features

S. No.	*Location*	*Soil Type*	*Max_ Temperature*	*Min_ Temperature*	*Rainfall*	*Humidity*
1	Chefe Donsa	Black Soil	31.0	13.0	811.2	68.0
2	Chefe Donsa	Black Soil	30.5	16.0	811.2	72.0
3	Chefe Donsa	Black Soil	29.5	14.0	811.2	68.0
4	Chefe Donsa	Black Soil	30.0	13.0	811.2	54.0
5	Chefe Donsa	Black Soil	31.0	12.0	811.2	70.0

Standardization: Before building the model we need to standardize the features by using the method fit_transform(). Find the mean and standard deviation to show whether the features are in standardized format or not. The mean of the dataset is -0.0 and the standard deviation of the dataset is 1.0.

5.3.1 Splitting the Dataset into Training and Test Sets

The considered dataset was split into the training set with 80% of the data and the test set with 20% of the data. The training set contains 6461 rows for 11 independent variables and 6461 rows for one dependent variable. The test set contains 1616 rows for 11 independent variables and 1616 rows for one dependent variable.

5.3.2 Model Building

The models were built by using the Random Forest Regressor, Ada Boost Regressor, Gradient Boosting Regressor, and Xgboost algorithms.

5.3.3 Model Building with Random Forest Regressor

The execution time for fitting the Random Forest Regressor algorithm by using the training data is 1.03226 Sec. Predictions were made by using the test dataset. Mean absolute error, mean squared error, and root mean

Table 5.6 The Transformed Dataset

SNo	*Max_ Temperature*	*Min_ Temperature*	*Rainfall*	*Humidity*	*Location_ Alemtena*	*Location_ Chefe Donsa*	*Location_ DZ-BS*	*Location_ DZ-LS*	*Location_ Denkaka*	*Location_ Minjar*	*SoilType_ Light Soil*
0	24.0	14.5	543.8	48.0	0	0	0	0	0	0	0
1	22.0	13.0	811.2	69.0	0	1	0	0	0	0	0
2	25.0	15.0	510.9	60.0	0	0	0	0	1	0	0
3	27.0	13.0	543.8	65.0	1	0	0	0	0	0	1

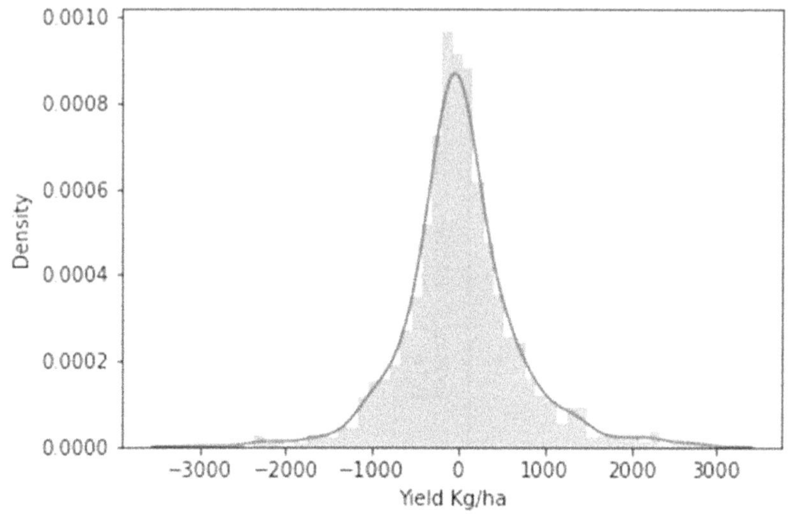

Figure 5.1 sns.distplot(y_test,y_pred).

squared error are 465.43152, 449065.76721 and 670.12369 respectively. Model accuracy for the test dataset and training dataset are 0.63316 and 0.94373 respectively.

The bell-shaped Figure 5.1 shows the Yield Kg/ha test dataset and the predicted density dataset. Figure 5.2 shows the scattered graph.

5.3.4 Model Building with Ada Boost Regressor

The execution time for fitting the Ada Boost Regressor algorithm by using the training data is 0.86647 Sec. Predictions were made by using the test dataset. Mean Absolute error, mean squared error, and root mean squared error are 610.07187, 589747.06235, and 767.94990 respectively. Model accuracy for the test dataset and training dataset are 0.51824 and 0.529473 respectively.

The bell shaped Figure 5.3 shows the Yield Kg/ha test dataset and the predicted Density dataset.

Figure 5.4 shows the scattered graph for the test dataset and the predicted dataset.

5.3.5 Model Building with Gradient Boosting Regressor

The execution time for fitting the Ada Boost Regressor algorithm by using the training data is 1.15264 Sec. Predictions were made by using the test

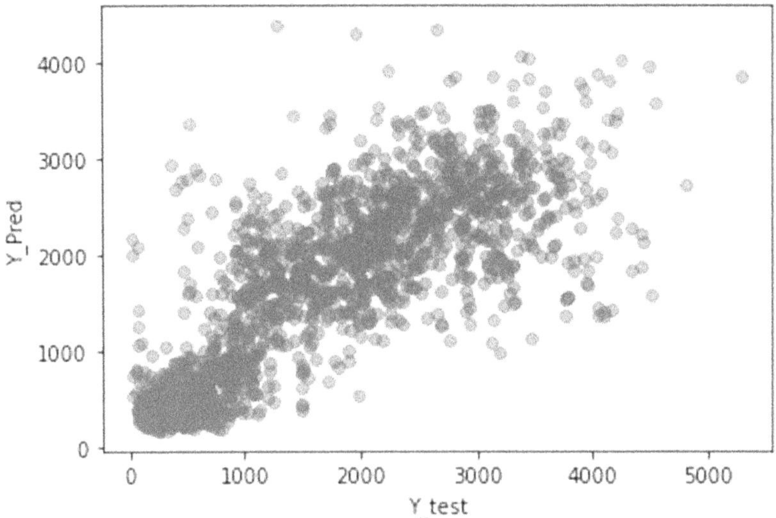

Figure 5.2 plt.scatter(y_test,y_pred, alpha=0.5).

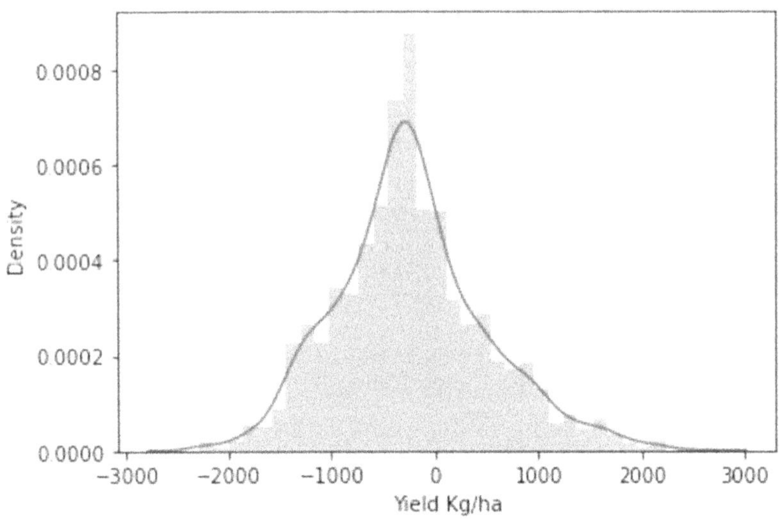

Figure 5.3 distplot(y_test,y_pred).

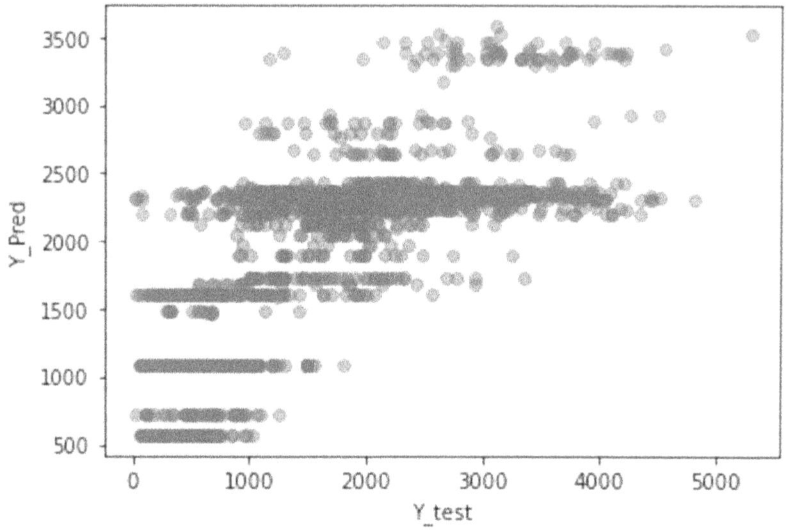

Figure 5.4 scatter(y_test,y_pred,alpha=0.5).

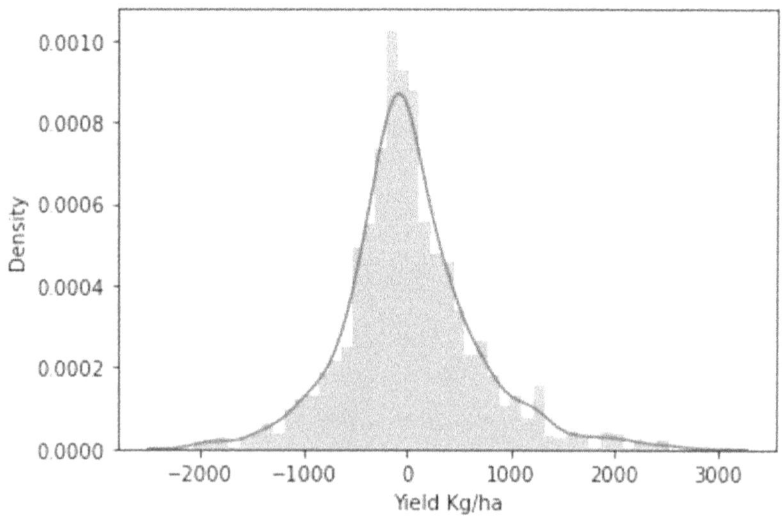

Figure 5.5 sns.distplot(y_test,y_pred).

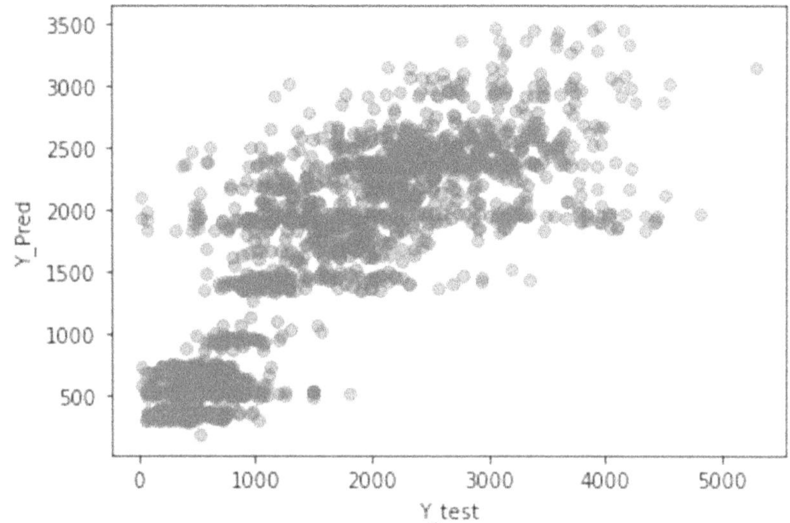

Figure 5.6 plt.scatter(y_test,y_pred,alpha=0.5).

dataset. Mean absolute error, mean squared error, root mean squared error are 459.31197, 410940.56675 and 641.04646 respectively. Model accuracy for the test dataset and training dataset are 0.66430 and 0.68863 respectively.

The bell-shaped Figure 5.5 shows the Yield Kg/ha test dataset and the predicted density dataset.

Figure 5.6 shows the scattered graph for the test dataset and the predicted dataset.

5.3.6 Model Building with Xgboost

The execution time for fitting the Xgboost algorithm by using the training data is 0.26891 sec. Predictions were made by using the test dataset. Mean absolute error, mean squared error, and root mean squared error are 464.52704, 432016.47275, and 657.27960 respectively. Model accuracy for the test dataset and training dataset are 0.64709 and 0.84768 respectively.

The bell-shaped Figure 5.7 shows the Yield Kg/ha test dataset and the predicted density dataset.

Figure 5.8 shows the scattered graph for the test dataset and the predicted dataset.

Table 5.7 shows the mean absolute error, mean squared error, root mean squared error. Accuracy for the test dataset and accuracy for the training dataset are 465.43152, 449065.76721, 670.12369, 0.63316, 0.94373

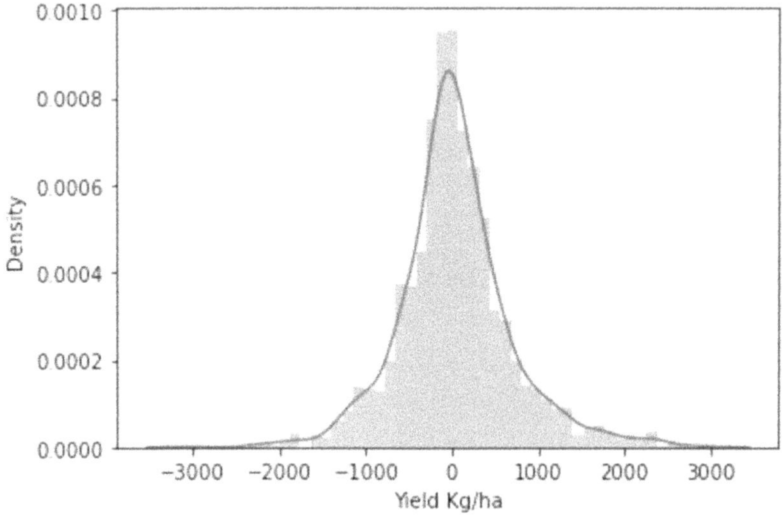

Figure 5.7 sns.distplot(y_test,y_pred).

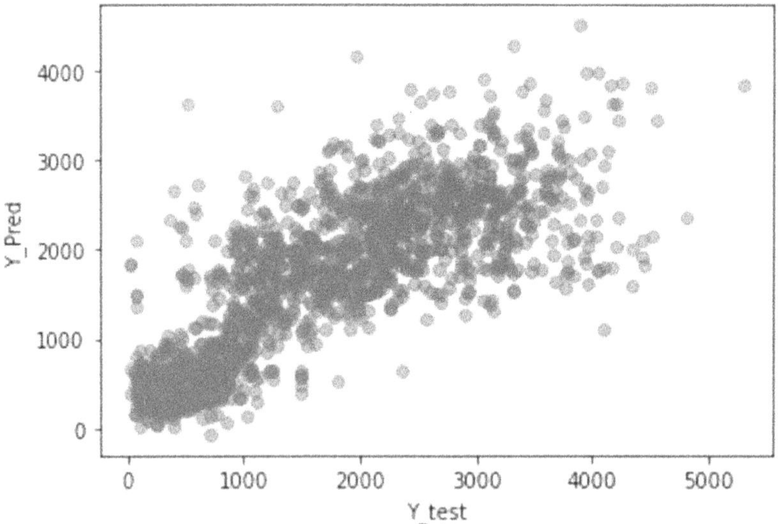

Figure 5.8 plt.scatter(y_test,y_pred, alpha=0.5).

Table 5.7 Empirical Evaluation of Classification Methods

S. No.	*Algorithm*	*MAE*	*MSE*	*RMSE*	*Model accuracy for the test dataset*	*Model accuracy for the training dataset*
1	Random Forest Regressor	465.43152	449065.76721	670.12369	0.63316	0.94373
2	Ada Boost Regressor	610.07187	589747.06235	767.94990	0.51824	0.52947
3	Gradient Boosting Regressor	459.31197	410940.56675	641.04646	0.66430	0.68863
4	Xgboost	464.52704	432016.47275	657.27960	0.64709	0.84768

Table 5.8 Comparison table of each classifier based on Time

S. No.	*Algorithm*	*Execution Time (Sec.)*
1	Random Forest Regressor	1.03226
2	Ada Boost Regressor	0.86647
3	Gradient Boosting Regressor	1.15264
4	Xgboost	0.26891

for Random Forest Regressor; 610.07187, 589747.06235, 767.94990, 0.51824, 0.52947 for Ada Boost Regressor; 459.31197, 410940.56675, 641.04646, 0.66430, 0.68863 for Gradient Boosting Regressor; 464.52704, 432016.47275, 657.27960, 0.64709, 0.84768 for Xgboost algorithms respectively. Each algorithm has its performance in terms of various considered performance measures. The Gradient Boosting Regressor algorithm has less Mean Absolute Error (MAE), Mean Squared Error (MSE), and Root Mean Squared Error (RMSE) compared with Random Forest Regressor, Ada Boost Regressor, and Xgboost algorithms. The Gradient Boosting Regressor algorithm has more accuracy for the considered test dataset compared with the other algorithms, whereas the Random Forest Regressor has more accuracy for the considered training dataset compared with the other algorithms such as Ada Boost Regressor, Gradient Boosting Regressor, and Xgboost algorithms.

Table 5.8 shows the execution times (sec.) of Random Forest Regressor; Ada Boost Regressor; Gradient Boosting Regressor; and Xgboost algorithms are 1.03226, 0.86647, 1.15264, 0.26891 respectively. The experimentation shows that the Xgboost algorithm is better in terms of execution time.

22. Liakos, K. G., Busato, P., Moshou, D., Pearson, S., & Bochtis, D. (2018). Machine learning in agriculture: A review. *Sensors*, 18(8), 2674.
23. Lobell, D. B., & Burke, M. B. (2010). On the use of statistical models to predict crop yield responses to climate change. *Agricultural and Forest Meteorology*, 150(11), 1443–1452.
24. Lopes, A. R., & Garcia, N. M. (2019). Data-driven decision making in agriculture: A systematic literature review. *Computers and Electronics in Agriculture*, 156, 417–431.
25. Miao, Y., Li, M., Cao, H., Ma, Z., Yang, L., & Li, G. (2020). Applications of machine learning in agriculture: A systematic review. *Computers and Electronics in Agriculture*, 176, 105681.
26. Mishra, A., & Patel, S. (2020). Applications of machine learning in agriculture: A review. *International Journal of Engineering and Advanced Technology (IJEAT)*, 9(6), 2854–2858.
27. Kamilaris, A., & Prenafeta-Boldú, F. X. (2018). Deep learning in agriculture: A survey. *Computers and Electronics in Agriculture*, 147, 70–90.
28. Krizhevsky, A., Sutskever, I., & Hinton, G. E. (2012). ImageNet classification with deep convolutional neural networks. In *Advances in Neural Information Processing Systems* (pp. 1097–1105), Editors: F. Pereira, C. J. C. Burges, L. Bottou, K. Q. Weinberger, Publisher: Curran Associates Inc., USA.
29. Kussul, N., Lavreniuk, M., Skakun, S., & Shelestov, A. (2017). Deep learning classification of land cover and crop types using remote sensing data. *IEEE Geoscience and Remote Sensing Letters*, 14(5), 778–782.
30. Mohanty, S. P., Hughes, D. P., & Salathé, M. (2016). Using deep learning for image-based plant disease detection. *Frontiers in Plant Science*, 7, 1419.
31. Mou, L., Zhang, L., Shen, Y., Cui, J., & Zhang, H. (2019). A review of deep learning methods for object detection in agricultural applications. *Computers and Electronics in Agriculture*, 165, 104943.
32. Raza, S., Zhang, Z., Farooq, U., & Liu, L. (2019). Machine learning for agricultural big data: A survey. *ACM Computing Surveys (CSUR)*, 52(5), 1–37.
33. Zhang, Y., Hu, Y., Zhu, D., & Chen, D. (2020). Deep learning for image-based plant disease detection: A review. *Frontiers in Plant Science*, 11, 558642.
34. Zhang, N., Wang, Y., Liu, Y., & Xie, X. (2020). Deep learning for remote sensing image analysis: A technical tutorial on the state of the art. *IEEE Geoscience and Remote Sensing Magazine*, 8(4), 32–58.
35. Sugiura, R., Komatsu, T., Yanai, K., & Nakano, K. (2019). A review of machine learning in agriculture. *Computers and Electronics in Agriculture*, 163, 104859.
36. Yu, M., Zhu, X., & Liu, H. (2020). Machine learning for crop disease diagnosis: Literature review and challenges. *Computers and Electronics in Agriculture*, 178, 105753.

Chapter 6

The Role of Chatbots in the Agriculture Sector

Partha Ghosh and Suradhuni Ghosh

6.1 INTRODUCTION

Agriculture is the primary source of livelihood for most citizens, providing food, employment, income, and raw materials for manufacturing. However, in some areas, agriculture is limited due to a lack of information and knowledge on modern farming technologies. Farmers in rural and sub-urban areas struggle to access agricultural knowledge and real-time information on the latest practices, leading to poor crop productivity.

A chatbot can provide useful answers, advice, and recommendations on specific farming concerns, providing a platform for the government to reach out to farmers and obtain feedback on governance through agricultural services. The use of chatbots has also increased to solve farmers' problems.

Agriculture, the oldest and most crucial profession, has evolved to increase crop yields and adapt to resource scarcity. With an estimated $5 trillion global industry, it contributes significantly to the global economy. Artificial Intelligence (AI) is being used to enhance efficiency and improve yields. The industry is also utilizing chatbots for immediate solutions to farmers' problems, with machine learning enabling them to analyze soil issues and provide recommendations for different land types. Trace Genomics is helping farmers analyze potential problems using ML. Research is being conducted to develop the best farm-bot to address labour shortages and transition from an agrarian society to a more sustainable one.

Chatbots offer 24/7 customer service, freeing up phone lines and being less expensive than hiring human support staff. They use AI and natural language processing to understand customer needs and provide the necessary help. Companies like chatbots because they can collect data on queries, response times, and satisfaction. However, chatbots have limitations, such as not fully comprehending customer input, providing incoherent answers, and being limited in their scope of queries. This can lead to customer dissatisfaction with generic feedback and can be expensive to implement and maintain, especially if they need frequent customization and updates.

DOI: 10.1201/9781003676102-6

The global population is expected to grow, increasing the need for food. AI and ML are expected to strengthen and improve the agriculture industry, even as people move to cities for jobs. Technology will ensure food production meets global requirements, even as people shift from farms to urban areas.

6.2 HOW AI CAN HELP THE AGRICULTURE SECTOR TO OVERCOME THE MOST CHALLENGING HURDLES

Due to the fast population development and land constraint, technology has changed the agriculture industry, allowing farmers to enhance production and nurture crops more effectively, changing the face of farming.

6.2.1 Scope of AI in Agriculture

Global agricultural output is expected to rise by 70% by 2050, according to the UN Food and Agriculture Estimate. The $5 trillion global agriculture sector is concentrating on Artificial Intelligence (AI) technology to enhance a range of agriculture-related tasks, hence promoting innovation, expansion, and influence in the food chain, in order to satisfy this need.

6.2.2 AI and Its Importance in the Agriculture Industry

Analysing farm data: Farmers used to have to physically manage a variety of duties, including weather forecasting, soil monitoring, and insect control. AI has made it simpler to forecast real-time aspects including weather, temperature, water use, and soil conditions, enabling improved decision-making. With the use of this technology, farmers may increase yields by making better decisions from the planning stage of farming.

Precision Farming: By using data analysis to identify pests, plant diseases, and undernourished crops, AI-based solutions improve precision agriculture by raising agricultural accuracy and quality.

Detecting weeds and toxins: To prevent excessive toxicity and to promote the production of nutritious food, sensors play a critical role in identifying and removing weeds as well as selecting the right herbicides.

Better forecasting: Farmers can foresee weather patterns and other factors impacting crop output with the help of AI's predictive algorithms. In small farms with little data and experience, these models are essential because they save financial waste and guesswork. For these farms to continue operating and to contribute to the agricultural economy, which ensures revenue for the farming community, AI deployment is imperative.

Monitoring: Farmers can now monitor their fields more professionally thanks to AI technology, and drone data collection increases productivity. In less time, these drones can pinpoint trouble spots in bigger fields that are hard for people to keep an eye on.

6.3 ADDRESSING THE SCARCITY OF LABOUR

- Automation: Because more people are migrating to cities, there is a labour shortage in agriculture, a labour-intensive industry. AI technology is automating business processes and helps farmers with a range of duties, including spotting trouble spots in the field and harvesting crops at a larger volume. AI bots can facilitate this change.
- Dissemination of information: Agritech startups are using AI chatbots to provide farmers with technical support. These chatbots respond to queries and give recommendations and advice based on farmers' current agricultural methods.
- Cost efficiency: Using technology to automate repetitive and time-consuming processes in agriculture increases cost efficiency.
- Lowering labour expenses and labour costs: Farmers can satisfy the world's food demand more effectively and productively by doing away with tedious procedures, cutting expenses, and minimizing mistakes in crop monitoring.

Hence, there will always be a need to cultivate food. Actually, it will rise in line with the demands of the expanding population. Thankfully, farmers will be able to run larger farms more efficiently and economically because of artificial intelligence. With the use of cognitive technology, they can produce food more efficiently.

6.4 WHAT IS CHATBOT AND ITS BENEFITS?

An automated computer software called a chatbot uses human speech recognition to answer client questions. Contemporary chatbots converse with clients in the same way as they would a human agent by utilizing AI/ML and natural language processing. If necessary, they may elevate routine inquiries to human agents and address them with efficiency.

One of the chatbots' advantages is that they can add a personal touch to communication. They chat with clients naturally and offer an interactive one-on-one experience. They can also provide personalized product and service recommendations based on the visitors' responses.

Businesses may use chatbots to automate processes like marketing, sales, and customer service. Additionally, they may assist companies in comprehending how clients engage with their chatbots. Additionally, chatbots are always available, meaning that when real humans aren't, they can still engage with site visitors and potential consumers. They can easily assist consumers in finding the right pages or links to utilize farmers' sites correctly and provide straightforward answers to basic inquiries. Ultimately, this can help save time and money.

6.5 AI AND CHATBOTS IN AGRICULTURE

The oldest occupation, agriculture, has changed throughout the ages as a result of resource constraint and population increase. The quality of agricultural goods has increased thanks to artificial intelligence and machine learning, and chatbots offer immediate support. These advanced instruments are being employed in drones, intelligent monitoring systems, and agricultural robotics. Ultra-precision agriculture is being made possible by artificial intelligence, which has several advantages for farmers.

Let us examine some advantages of AI and chatbots in the field of agriculture:

1. Recognition of Images: A key component of machine learning is image recognition, which helps with crop identification, crop health detection, pesticide and fertilizer usage, assessment of vitamin shortages, and overall crop quality evaluation.
2. Assistance from Chatbots: Trained with AI and ML, chatbots provide farmers with real-time support, translating across all languages and putting them in touch with knowledgeable specialists.
3. Soil Analysis: AI-driven technology looks at the land's base and stops soil deterioration to help assess soil quality, pinpoint problems, and offer tailored solutions to enhance agricultural operations.
4. Drone-based Crop Analysis: Drones are used in agriculture to monitor fields and carry out spraying duties. They are also used for irrigation, wildlife monitoring, disaster relief, and crop health evaluation.
5. Forecasting the Weather: Accurate weather forecasts made possible by artificial intelligence help farmers plan their crops according to rainfall, humidity, and wind speeds.

Enhancing data science, monitoring, and processing, chatbots, AI, and ML have completely changed the agriculture industry. Commission agents in Mandi accounting now have more work to do in addition to an increase in agricultural products. To solve these issues and offer possibilities, Bijak, the top agricultural trade app in India, created ChargeERP.

6.5.1 Chatbots for Farmers

Chatbots, such as Microsoft's **FarmVibes.Bot**, function as a mediator between farmers and their clientele, offering support with product queries, inventory verifications, and database administration. They serve as online farm assistants, offering individualized guidance and up-to-date information on the climate, agricultural markets, and other relevant topics. Currently in use by more than 500,000 farmers in sub-Saharan Africa, chatbots are an invaluable agricultural tool.

6.5.2 Using AI-powered Chatbots, Kissan GPT is Transforming Indian Agriculture

The improved capabilities of AI are causing a disruption in several industries, including the agriculture industry.

Based on ChatGPT, Kissan GPT is an AI-powered chatbot designed to help Indian farmers increase agricultural output. It offers real-time advice in different languages. Indian farmers may increase their agricultural production with the aid of Kissan GPT, an AI-powered chatbot. This chatbot is built on two AI chatbots that provide replies in a conversational manner: ChatGPT and Whisper.

6.5.2.1 *What is Kissan GPT?*

Introduced by Pratik Desai (on March 15, 2023), Kissan GPT, as in Figure 6.1, is an AI-powered chatbot that offers real-time advice on crop farming, irrigation, and pest control with the goal of assisting India's neglected agricultural sector and maximizing farmers' output.

Kissan GPT is a brand-new, 'powerful' AI chatbot that can assist Indian farmers in real-time with their agricultural questions and offer advice on crop production and other topics.

Figure 6.1 Kissan GPT is an AI-powered Chatbot.

6.5.2.2 *Kissan GPT Interface*

With Kissan GPT, farmers can speak with an AI chatbot in ten Indian languages—including Spanish, Japanese, Portuguese, and Indonesian—and get thorough answers in a matter of seconds by using the microphone on their phone or laptop. Kissan GPT Interfaces are shown in Figure 6.2.

6.5.2.3 *How Does Kissan GPT Work?*

With the use of GPT-4 and customized embeddings, Kissan GPT—which is aimed at farmers, students, researchers, and enthusiasts—will hopefully improve with time. It now employs OpenAI's ChatGPT technology and knowledge base to provide replies.

6.5.2.4 *What's Next for Kissan GPT?*

Kissan GPT, an AI chatbot, will handle 10% of farmer call center calls, according to Pratik Desai, who also stated that the chatbot is looking to work with government and agricultural institutions. Languages and mobile apps are also being developed.

6.5.3 How Kissan GPT Increases Revenue for Indian Farmers

6.5.3.1 *The Difficulties that the Farming Community of Indian Faces*

The Indian farming community faces difficulties as a result of knowledge gaps and limited access to developing technology. Since farming employs more than 60% of the workforce and accounts for 18% of India's GDP, the industry must be protected. Easy-to-use chatbots such as Kissan GPT have the potential to close this knowledge gap and transform India's agriculture industry.

6.5.3.2 *How Does Kissan GPT Address the Problems that Indian Farmers Face?*

The problems encountered by Indian farmers are being addressed by Kissan GPT, an AI-powered voice assistant that offers knowledgeable agricultural advice in a variety of Indian languages. Nine Indian languages are supported on the platform: Hindi, Gujarati, Marathi, Tamil, Telugu, Kannada, Malayalam, Punjabi, and Bangla. In the future, Kissan GPT intends to expand its help to Assamese and Odia, tackling the linguistic accessibility issues that Indian farmers encounter.

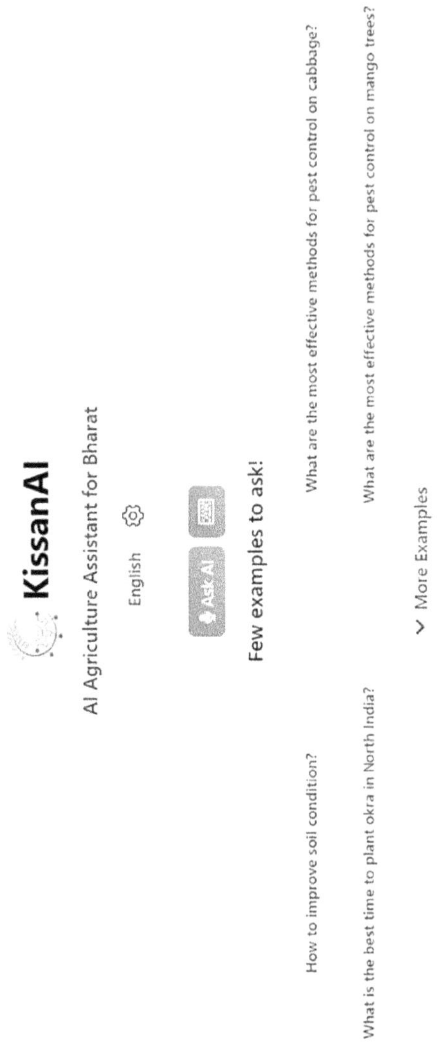

Figure 6.2 Kissan GPT Interface.

6.5.3.3 Effects and Prospects of Kissan GPT

Progressive farmers have expressed their satisfaction with Kissan GPT, a mobile app for farmers, and have even started organic groups to assist others in learning how to utilize the platform. The application has been used at night, which is not possible with conventional call-based methods. In a live session, more than 450 farmers took part, discussing their use cases and offering insightful commentary. Kissan GPT has a bright future ahead of it as its producers work to add new features, enhance accuracy, add more language support, and adjust to the changing needs of the agricultural industry.

6.5.3.4 Which App is Developed for Farmers in India?

Suvidha Kisan: One of the most popular farming applications in India, Kisan Suvidha was introduced by Prime Minister Narendra Modi in 2016. Additionally, the app gives farmers more control and information that might boost their crop productivity.

6.5.3.5 What is the Future Scope of Chatbot?

Future Chatbot challenges to be faced: Developing Natural Language Processing (NLP) skills for the chatbot is one of the biggest obstacles. This is essential for a chatbot to comprehend user requests and offer insightful answers.

6.5.3.6 What's the Name of the First AI Chatbot Designed Specifically for the Agriculture Industry in India?

The Ama KrushAI chatbot: Prof. Ganeshi Lal, the governor of Odisha, introduced
'Ama KrushAI,' India's first AI chatbot for the agricultural industry, during the concluding ceremony of 'Odisha 2023.'

6.5.3.7 How AI is Applied in Agriculture in India?

By encouraging the use of artificial intelligence and other technologies, the AI for Agriculture Innovation program is revolutionizing the Indian agriculture industry. The system is being used by more than 7,000 farmers for soil testing, quality control, and crop health monitoring.

6.5.3.8 A Smartphone Chatbot for Crop Farmers Powered by Machine Learning

The mobile chatbot using machine learning leverages a retrieval-based model with a domain-specific focus. A closed agriculture domain query is selected for execution. Figure 6.3 depicts the system architecture of the mobile chatbot for crop farmers that uses deep learning. Its main modules are as follows: sophisticated machine learning algorithms, intent recognition algorithm (query), knowledge base (dataset), Natural Language Processing (NLP), user interface and answer generator [1].

Farmers who cultivate cassava crops communicate with the system by sending inquiries and viewing the produced answers using the mobile device user interface. On the other hand, information gathered from recognized sources is kept in the knowledge base and generally utilized to assist machine learning algorithms in establishing guidelines for producing results after user query processing. Jacquard Similarity Algorithms and word shuffling were used to create classifiers for the users' intentions. In order to make it easier for the computer (chatbot) to understand user input, human natural language input was interpreted using NLP pre-processing techniques.

The system is divided into modules for user conversations, administrator duties, and adding, modifying, and deleting information. The system's architecture consists of several modules:

6.5.3.8.1 Natural Language Processing (NLP)

This module's goal is to analyze human natural language input and make it understandable to robots (chatbots). The following are the steps in NLP:

- Tokenization: The act of dividing the input phrases into a list of manageable chunks, or tokens, is known as tokenization. It is possible to tokenize paragraphs into sentences and sentences into words.

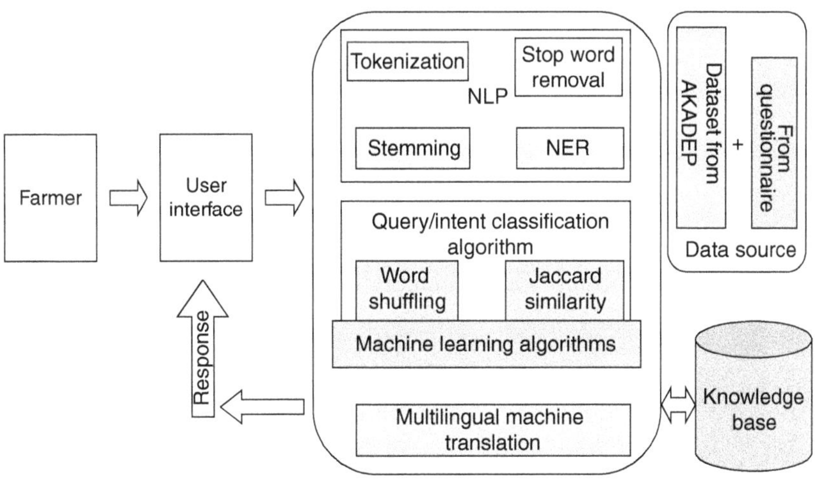

Figure 6.3 Architecture of ML-based Mobile Chatbot for Crop Farmers.

- Noise and stop word removal: Elimination of stop words and noise: Stop words, often known as noise, are a group of frequently used words in a language that don't convey any important information. English stop words include 'a,' 'the,' 'is,' 'are,' 'and,' and so on. The rationale behind eliminating stop words from text is to enable us to concentrate on the significant words.
- Stemming: The practice of distilling words to their stems or root variants is known as stemming. For example, 'walks,' 'walking,' and 'walker' may all be simplified to 'I walk' and 'computer,' 'computing,' and 'compute' to just 'comput.' As a result of treating words with the same stem consistently, this procedure may also improve the precision of the model.
- Named entity recognition: or NER, is the process of extracting named entities from a text, such as crop names. NER facilitates the identification and categorization of crucial information into predetermined categories, such as crop name, disease name, leaf color, and so forth, from inside more extensive unstructured text-based data.

6.5.3.8.2 Algorithm for Intent Recognition

The following machine learning methods were taken into consideration in order to create the classifier of the user intentions, or the objective the farmer has in mind while posing a query to the chatbot:

- The Jaccard Similarity Algorithm: An application for selecting and identifying intents in Python is the Jaccard Similarity method. It reduces the number of times a chatbot cannot ascertain the intents of a user and improves intent identification accuracy. The solution with the highest similarity score is retrieved by the algorithm, as shown in Figure 6.4. Improving chatbot answers requires the use of this technique.

The following is a summary of the primary steps in the system:

i. A chatbot uses the User Interface (UI) to get the farmer's inquiry.
ii. Forward the query to the NLP for named entity recognition, tokenization, stop word elimination, and stemming.
iii. By deriving the purpose from the user's message, the machine learning module utilizes the structured input from Natural Language Processing (NLP) to forecast the optimal course of action.
iv. The chatbot database receives a notice to read the farmer's query.
v. Determine the Jaccard Index between the database's questions and the user's query.

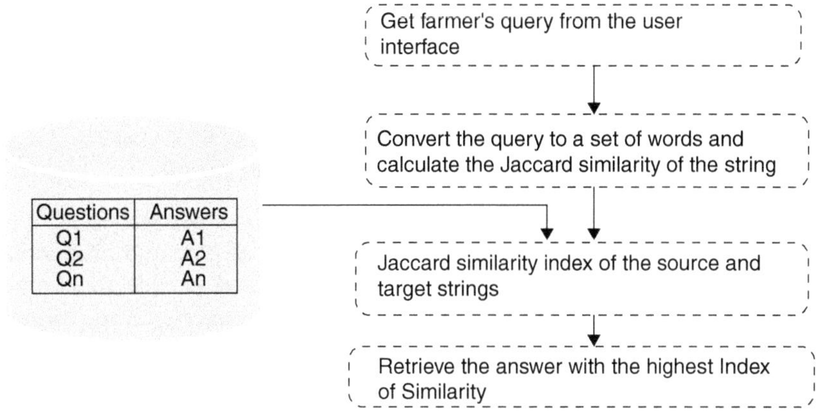

Figure 6.4 Jaccard Similarity Methods.

Table 6.1 Sample Application of Word Shuffling Technique

Before word shuffling	*After word shuffling*
Which kind of soil works best for cassava cultivation?	• Which kind of soil is ideal for cassava cultivation • Which kind of soil is ideal for growing cassava? • Which kind of soil is ideal for growing cassava?
Which month is ideal for cassava planting?	• Which month is ideal for cassava planting • Which month is the best to cultivate cassava? • Which month is best to cultivate cassava?

vi. Using the input text's Jaccard Similarity Index and the predetermined text, the response generator finds the answer to the question that comes the closest.

vii. The notice 'I don't understand your question' or 'Please rephrase it' appears if the user's purpose cannot be detected.

- Word shuffling: The study employs word shuffling to expand training data, ensuring grammatical correctness and sentence meaning. Each sentence is tokenized into separate words, shuffled, and gathered to create a new sentence. The resulting sentences are validated to maintain grammatical correctness. The validated sentences are then used to extend the initial dataset, ensuring wider coverage for user intents. When applied to input parts of the dataset, grammatical correctness may not be necessary, as users may provide incorrect inputs. In such cases, word shuffling can be beneficial for training chatbots. An example is shown in **Table 6.1**.

6.5.3.8.3 Machine Translation in Multiple Languages

When replying to a farmer's request, the framework aims to implement a multilingual translation from the farmer's native tongue to English and vice versa.

6.6 CHATGPT-4: USING AI TO BOOST FOOD SECURITY AND FOOD PRODUCTION IN AGRICULTURE

GPT-4 and other artificial intelligence chatbots have become more well-known due to their capacity to maximize farmers' income and production. OpenAI's GPT-4 natural language processing model is able to comprehend human language and give precise answers to queries from farmers. Enhancing the accuracy of crop and soil analyses, staying current with industry trends and technologies, automating repetitive tasks like crop monitoring and pest control, and optimizing decision-making by accessing real-time data on weather, soil conditions, and pest infestations are just a few of its potential advantages.

Through the analysis of data from several sources, GPT-4 not only warns farmers about possible issues or possibilities but also offers them insights into new goods and practices. The potential advantages of GPT-4 will only increase with further technological advancement, completely changing the agriculture sector.

6.6.1 Increasing Crop Yields using GPT-4: Investigating AI's Potential in Agriculture

In order to generate larger yields with less resources, efficient agricultural methods are required due to the growing world population. Crop management and agriculture might undergo a radical transformation thanks to Artificial Intelligence (AI). With the help of recent developments in natural language processing, Generative Pre-trained Transformer (GPT-4) is a potent artificial intelligence tool that can produce text in response to a request. In order to increase agricultural yields, GPT-4 is being evaluated through data analysis, model development that predicts soil types and environmental conditions, effective planting and harvesting techniques, and appropriate water, fertilizer, and other input amounts. Farmers may utilize GPT-4 to assist in the making of well-informed decisions about when and how to apply best practices for their crops by integrating this data with real-time meteorological and soil data.

In addition to providing crop-specific predictions and best practices for anticipating and responding to extreme weather events, GPT-4 may create predictive models to assist farmers in anticipating and planning for extreme weather occurrences. By providing fresh insights from massive databases, it may also transform agricultural research by assisting scientists in creating innovative approaches to increase crop yields and sustainability. Before the technology is widely used, there are a number of issues that need to be resolved, including testing and improving it, locating and assessing data sources, and taking ethical issues like possible bias in decision-making into account.

6.6.2 GPT-4 Focuses on the Use of AI-powered Chatbots in Farming to Enhance Efficiency

Although it might be difficult, farming is an essential activity that produces food for everyone on the planet. An increasing number of farmers are using artificial intelligence to ease their workloads as technology progresses.

Artificial Intelligence (AI)-driven chatbots are computer programs that leverage natural language processing to facilitate dialogues with users. They help farmers in managing inventory, forecasting the weather, and choosing crops. These chatbots become more knowledgeable about user preferences and agricultural circumstances as they gain experience, which eventually increases their efficiency and dependability.

Food merchants and agricultural enterprises may also benefit from their expertise on supply chain management and customer service. More farmers will profit from AI-powered chatbots as technology develops, changing and streamlining agriculture.

6.6.3 Use of GPT-4 and AI-powered Chatbots to Mitigate Agricultural Losses

An AI-powered chatbot system has been created by a group of scientists from the Universities of California, Berkeley and Southern California with the goal of lowering agricultural losses. With the use of a GPT-4 language model, the system was trained to give farmers early information about possible agricultural dangers like insect problems or harsh weather. By giving farmers up-to-date information on crop illnesses, insect infestations, and weather occurrences, the team evaluated the system. The findings revealed a loss reduction of 5.5% on average, with a high of almost 16%. According to the study, chatbots powered by AI might be a useful tool for farmers looking to lower their agricultural losses.

The group proposes using these chatbots to track and forecast the development of diseases, weeds, and pests in crops. The results demonstrate how AI-powered chatbots may assist farmers in lowering losses and increasing yields.

6.6.4 AI-powered Conversational Interfaces with GPT-4 to Increase Agricultural Productivity

In order to increase productivity, boost crop yields, and lower costs of production, the agricultural sector is using Artificial Intelligence (AI) more and more in the form of robots and machine learning. The potential of GPT-4, an enhanced variant of OpenAI's GPT-3, in agricultural applications is being investigated. A GPT-4-based conversational interface has been built by researchers at the University of California, San Diego (UCSD) to answer inquiries from farmers on crop development, insect management, and other agricultural subjects. The approach has been proven to be more efficient than conventional techniques, accurate, quicker, more dependable, and less expensive because it doesn't require extra staff or training.

Researchers at UCSD are also looking at further applications of GPT-4 in agriculture, including precision agriculture systems development, soil health, insect population monitoring, and crop field monitoring. If realized, GPT-4's potential might completely transform farming techniques and result in more affordable and productive agricultural production techniques, which would increase global food security and economic prosperity.

6.7 SUBSTANTIAL IMPACT OF CHATGPT AND BLOCKCHAIN TECHNOLOGY ON THE AGRICULTURE INDUSTRY

ChatGPT enables farmers to sell their products directly to consumers, while Blockchain technology is gaining popularity in the agricultural sector due to its numerous applications, promoting transparency and security in the industry.

6.7.1 What is the Technology Behind Blockchains?

With smart contracts and Blockchain technology, farms can now maintain safe records in a novel way. By allowing farmers to sell directly to customers, it lowers supply chain fraud and corruption. The unchangeable nature of Blockchain guarantees confidence and food safety verification. It tackles difficulties including food safety traceability problems and commodities market transparency problems. Overall, farmers can manage their supply chains more effectively and safely with the use of Blockchain technology.

6.7.2 What is ChatGPT?

One well-known Artificial Intelligence (AI) tool created by Open AI, to produce replies to prompts that resemble those of a person, is called ChatGPT. Machines may now converse more smoothly with humans thanks to deep learning algorithms that grasp the context of conversations. Customer support, e-commerce, and education are just a few of the uses for which ChatGPT's adaptability makes it perfect.

But possible chatbot biases—especially against underrepresented communities—raise ethical questions.

The potential influence of ChatGPT on agriculture is substantial, notwithstanding these reservations, as growers look for creative methods to boost yields and save expenses.

6.7.3 What Is the Impact of ChatGPT on Agriculture?

Farming might undergo a transformation thanks to ChatGPT, an AI-powered chatbot that gives farmers real-time access to data on weather patterns, market trends, and crop management. Farmers can more easily comprehend complicated agricultural concepts and practices because of its capacity to communicate in local languages. ChatGPT analyzes data from sources such as satellite photography and farm sensors to maximize resource allocation and minimize waste, enabling farmers to make well-informed decisions regarding their crops. Reduced risks from pests and the effects of climate change are achieved by early detection of illnesses or pests harming crops.

An additional feature of ChatGPT is semi-simple stakeholder contact using chatbots or other conversational interfaces. Along the whole value chain, automation of manual processes raises profitability because it increases productivity and lowers labour costs. A number of uses for ChatGPT-powered chatbots in agriculture include crop planting based on soil and weather patterns, as well as helping with customer care questions about farming equipment.

ChatGPT has the power to revolutionize agriculture by giving smallholder farmers better access to global decision-making tools and efficiencies. Through simple mobile device interaction with the chatbot, farmers without access to traditional extension services may receive timely guidance without having to drive great distances or wait for extra personnel to arrive.

6.7.4 What is the Primary Function of Blockchain Technology in the Agricultural Sector Revolution?

The ability of Blockchain technology to provide safe and transparent record-keeping and transaction processes is transforming the agriculture industry. By enabling farmers to follow their products from farm to table, it enhances supply chain management and lowers waste while enhancing food safety.

Blockchain's smart contracts facilitate easier financial transactions between suppliers and farmers. Because Blockchain is decentralized, it does not require middlemen, which results in cheaper transaction costs and more effective trading methods. The traceability benefits of Blockchain records' openness enable customer trust in food systems and lessen fraud risks associated with counterfeiting operations. All things considered, Blockchain has great promise for revolutionizing the agriculture industry.

6.7.5 What Is ChatGPT's Principal Function in the Agriculture Industry?

A sophisticated artificial intelligence language model called ChatGPT, or Conversational GPT, may produce replies in natural language conversations that resemble those of a person. It may facilitate better communication between farmers and other agricultural stakeholders, offer tailored guidance depending on particular soil types or meteorological conditions, and help agriculture experts with research by giving them rapid access to pertinent data.

To put this technology into practice, industry professionals and AI developers must work together to design workable solutions that are customized for various agricultural scenarios. When incorporating AI models like ChatGPT into agriculture, there is still more unrealized potential.

6.7.6 What Effects Can Blockchain Technology Have on Agriculture?

Agriculture is significantly impacted by Blockchain technology, a decentralized digital ledger. It encourages sustainable practices by enabling farmers to document data on crop yields and soil health over time, improving product transparency and traceability from farm to table. To ensure proper tracking of yields, costs, and revenue, farmers may save their farming data securely on an immutable ledger. Additionally, Blockchain gives the supply chain more control, guaranteeing that goods are delivered to customers effectively and safely.

Utilizing smart contracts to store data about seed origin and pesticide usage on an irreversible ledger, producers can demonstrate the legitimacy of their organic goods. Because Blockchain technology increases responsibility amongst cross-functional teams participating in value chains and improves transparency among stakeholders within complicated networks, it provides prospects for revolutionizing a number of industries, including agriculture.

6.7.7 How Can ChatGPT and Blockchain Technology be Applied to Agriculture?

- Agriculture might undergo a transformation thanks to ChatGPT and Blockchain technology, which can improve supply chain management and guarantee high-quality, contamination-free food.
- They can also be used in land registry to prove ownership and guard against property rights disputes. ChatGPT has applications in Natural Language Processing (NLP), pest management, crop disease diagnostics, and precision farming.
- Large volumes of agricultural data can be processed by them more efficiently than by people. When ChatGPT and Blockchain are used in tandem, the agricultural supply chain may be streamlined from planting to packing.
- Blockchain-based smart contracts automate transactions without the need for middlemen, offering transparency and automation.

6.7.8 ChatGPT Consequences for Agriculture

With ChatGPT's sophisticated artificial intelligence capabilities, massive volumes of data can be accurately and quickly analyzed, potentially revolutionizing the agricultural industry. This gives farmers information on the health of their crops, the weather, and the quality of the soil—all of which are essential for producing the highest possible yield.

In order to help farmers better prepare for the future of their crops without running the danger of suffering large losses from unanticipated events, ChatGPT may also help with predictive analysis by examining historical trends and projecting future results. Furthermore, ChatGPT is a useful tool for a variety of businesses since it may support precision farming techniques.

6.7.9 How ChatGPT Can Transform Agriculture

- By using predictive analytics algorithms to estimate crop yields based on past data, ChatGPT is transforming the agricultural industry. This enables farmers to maximize resources and prepare for next planting seasons.
- Drone technology and sensors can provide real-time notifications concerning abnormalities and problems. The conversational interface of ChatGPT makes it possible to automate manual processes like inventory management and maintenance in an effective manner.
- By integrating IoT devices with smart irrigation systems, the platform may save waste and boost agricultural output.

- Farmers may increase yield output by using precision agricultural techniques, such input customization, to assist them in making well-informed decisions regarding resource management.

6.7.10 Agricultural Applications of Blockchain Technology

With its ability to handle data securely and transparently throughout the supply chain, Blockchain technology has the potential to completely transform the agriculture sector. It can monitor the whole food production process, from the farm to the table, giving customers trustworthy information about where their food comes from. Real-time agricultural output and pricing tracking using Blockchain technology can also help farmers manage stocks more effectively.

Automation of agricultural payments based on preset conditions is possible with smart contracts enabled by Blockchain technology. Blockchain technology may also be used to build decentralized markets that will improve market accessibility and lower transaction costs for small-scale farmers.

6.7.11 How Will Blockchain Affect Agriculture in the Future?

By streamlining the supply chain, cutting down on transaction costs, and opening up peer-to-peer financing platforms to farmers, Blockchain technology has the potential to completely transform the agricultural industry. Enhancing crop yields, it can monitor crop growth conditions.

Through the automated use of smart contracts on decentralized networks, Blockchain technology may help guarantee fair pricing for farmers. An excellent tool for risk management in agriculture, it can also monitor weather trends and predict catastrophic occurrences such as floods or droughts.

By promoting sustainability, modern farming methods may be revolutionized, and consumer confidence in food systems could be strengthened.

6.8 FARMER'S COMPANION: INTERACTIVE AI BOT FOR INTELLIGENT FARMING

A conversational AI bot called Farmer's Friend seeks to assist farmers in using smart agriculture. It is a product of Digital Green, an organization that offers AI solutions to the farming industry. Questions about the weather, crops, soil, fertilizer, market prices, government programs, and other topics can be answered by Farmer's Friend. Additionally, it might offer recommendations and pointers for enhancing agricultural methods and

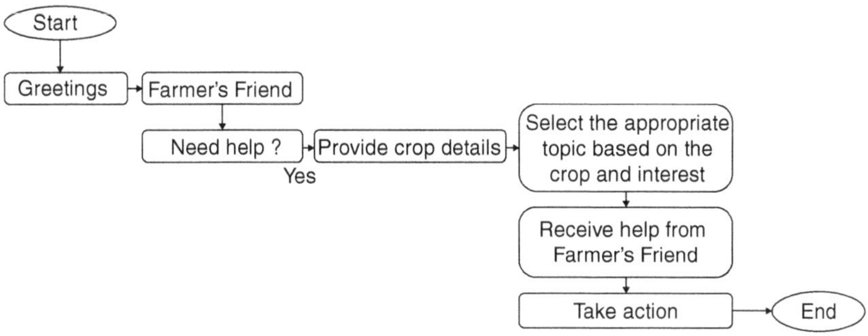

Figure 6.5 Diagram Representing a State Machine when a Farmer Starts Interaction with Farmers's Friend.

raising output. GPT-4, a cutting-edge natural language processing model that can produce thoughtful and pertinent replies, powers Farmer's Friend [2]. Figure 6.5 represents a state machine when a farmer starts interaction with Farmer's Friend.

Some of the features of Farmer's Friend are:

- English, Hindi, Tamil, Telugu, Kannada, Malayalam, Marathi, Gujarati, Bengali, Urdu, Punjabi, and more languages are supported.
- It can deliver precise and tailored responses by comprehending the context and intent of the user's inquiry.
- Over time, it can become more efficient by learning from the preferences and input of the user.
- To reach more farmers, it may integrate with other platforms like Facebook Messenger, WhatsApp, Telegram, and so forth.
- It is capable of producing graphical artworks associated with agriculture, including representations of crops, soil, and weather.

For farmers looking to expand their expertise in smart agriculture, Farmer's Friend is a creative and practical tool. By giving consumers trustworthy information and direction, it may help them save time and money. By optimizing their crop management, it may also assist them in lowering risks and raising revenues. For farmers who wish to cultivate more intelligent crops, Farmer's Friend is a helpful ally.

6.9 AGRICULTURE-SPECIFIC QUESTION ANSWER SYSTEM (AGRIBOT)

A group of researchers from the IIT Gandhinagar created AgriBot [3], a question-and-answer system tailored to the agricultural industry. Based on entity extraction techniques and a phrase embedding model, it provides answers to questions on weather, market pricing, plant protection, and government plans in India. Figure 6.6 depicts the working strategy of AgriBot.

The system is accessible round the clock via any electronic device, and the information is presented in an easily comprehensible manner. In addition to lessening the strain on the contact centre employees who supply this information, the system seeks to give farmers convenient access to reliable information regarding farming methods and trends.

6.10 ARTIFICIAL INTELLIGENCE IN INDIAN AGRICULTURE

AI and other technologies are revolutionizing the agriculture industry in India. More than 7,000 farmers use AI for soil testing, quality control, and crop health monitoring. AI-based solutions are being integrated by Indian agriculture technology businesses to address supply chain efficiency, soil fertility, crop yield, and predictive analytics. Agriculture may benefit from the incorporation of AI in areas like crop yield, soil fertility, and supply chain efficiency.

Artificial intelligence integration has the potential to greatly improve agriculture in a number of ways:

- In order to provide agricultural advisory services utilizing cognitive computing, drones for crop health monitoring, hydroponics for water and nutrient conservation, artificial lighting for ideal lighting conditions, and AI-powered cameras for pest detection and irrigation requirements, Microsoft has partnered with 175 farmers in Andhra Pradesh.
- These innovations seek to minimize energy use, cut carbon emissions, and increase productivity. Additionally, Microsoft provides agricultural, land, and fertilizer consultancy services in collaboration with 175 farmers. Drones are utilized for precision farming, insect identification, and crop health monitoring as they gather airborne photographs and data.

Artificial intelligence is anticipated to greatly accelerate the expansion of Indian agriculture by reducing stress and encouraging a move toward data-driven farming.

6.11 CNN-POWERED AGRICULTURE HELPER CHATBOT

One typical problem is the absence of pertinent knowledge on when and how to irrigate, plant seeds, and use pesticides. A multi-user chat application called AgroBot seeks to solve this by arming farmers with the information

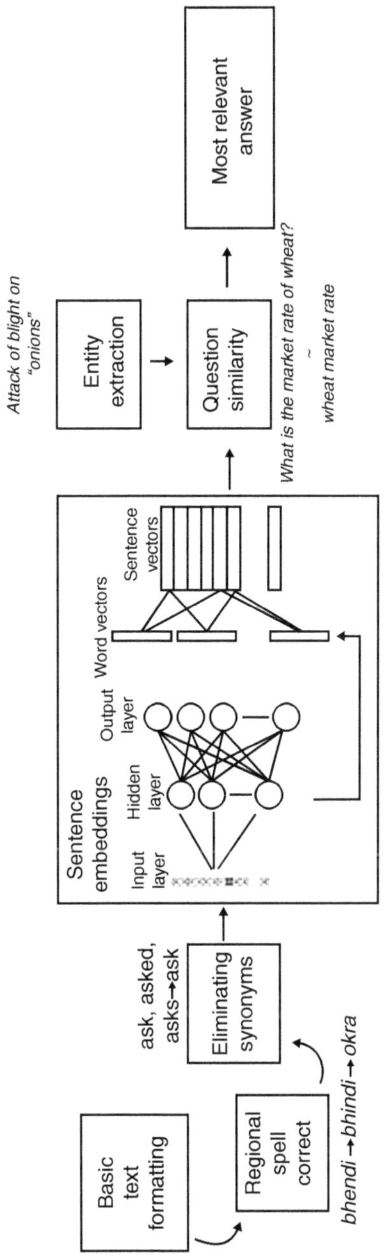

Figure 6.6 Working Strategy of AgriBot.

they need to prosper in a changing market. Using photos of crop leaves, the chatbot use a Deep Learning CNN algorithm to identify diseases and provide possible treatments. After that, farmers may ask the chatbot questions about their crops, such as what kind of crop they are growing, and it will answer with pertinent facts like soil type, rainfall, and other factors. The goal of this initiative is to increase Indian farmers' access to and productivity from farming [4].

Here, our goal is to create a chatbot that farmers can use to input photos of their crops and have a Deep Learning CNN algorithm identify problems and suggest treatments. In addition, the chatbot may display information about the crop, such as rainfall and soil type, and ask questions about it. The program interprets the farmer's inquiries using speech recognition technology. Figure 6.7 represents the flow chart.

1. Start: The user will use run to launch the application in this case.bat for output.
2. Upload Test picture: The test picture from the 'test images' folder will be uploaded in this module. It will predict the illness name depending on input.
3. Conversational AI: A chatbot automates consumer interactions between a business and its clients through conversational engagement. Thanks to machine learning techniques, chatbots that are powered by artificial intelligence are able to comprehend natural language and provide personalized responses to users. Retailers, travel agencies, media firms, insurance companies, and other industries were the main users of chat bots in the beginning. With the use of this new technology, agriculture might also help farmers discover the answers to their inquiries and offer suggestions and advice on matters pertaining to farming.
4. Exit: The end button in this module allows the user to exit the program.

Along with agricultural information, the messaging app will provide farmers access to specific data such as market pricing and weather forecasts. The AI-enabled technology will make contemporary farming more accessible to farmers faster, especially when it comes to comprehending the attitudes of Indian farmers and the structure of the agricultural industry.

6.12 HOW MIGHT CHATBOTS WITH ARTIFICIAL INTELLIGENCE ASSIST POLICYMAKERS?

Chatbots that are equipped with artificial intelligence have the potential to greatly aid policymakers. They can offer expedient access to pertinent

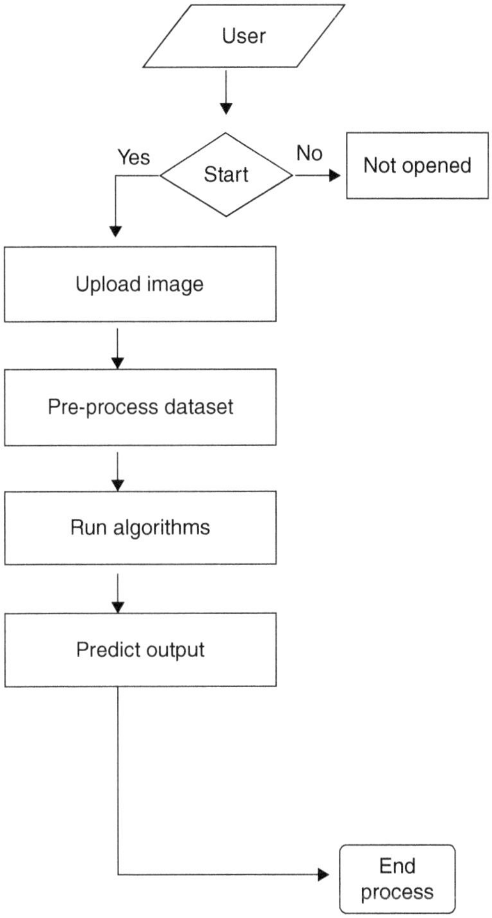

Figure 6.7 Flow Chart.

data and evidence from multiple sources, produce policy recommendations grounded in best practices and existing evidence, and improve communication and interaction with stakeholders. They must, however, ensure the accuracy and – dependability of the information they offer, comprehend the reasoning and logic underlying their conclusions, and assess the significance and efficacy of their conclusions.

Because of their training data or algorithms, chatbots may not be able to independently evaluate the reliability or correctness of the sources they utilize or quote, and they may provide biased or deceptive results. Additionally, they are unable to quantify or track how their suggestions influence policy outcomes or the well-being of the people they serve, and they lack a thorough grasp of their own outputs or the consequences of their recommendations.

Chatbots should be used in conjunction with other tools rather than as a replacement if they are to be effective. Using chatbot results should require policymakers to use critical thinking and judgment.

6.12.1 ChatGPT-based Chatbot to Assist Rural India

Rural India may benefit from the services of a chatbot powered by ChatGPT by

- Getting information on government programs, healthcare, education, and agriculture.
- It can close the divide between urban and rural communities by enhancing access to services like banking, insurance, and healthcare.
- The Bhashini Daan Mission is integrating new languages, and as a result, the chatbot supports many languages, increasing savings and financial awareness among rural populations.

Under the BHASHINI Mission, a local language translation program, MeitY is creating a WhatsApp-based chatbot to facilitate content creation in Indian languages and simple access to the internet and digital services in Indian languages, including voice-based access. The mission also seeks to provide Indian entrepreneurs, MSMEs, and public innovators with AI-based resources.

6.12.1.2 Actions to go Beyond Obstacles

- Establishing a digital public platform for Indian languages by the government.
- Lowering the digital divide in India's rural areas.
- Making inexpensive technology accessible.
- Lowering the resistance of behaviour to new technology.

Rural inhabitants' quality of life can be enhanced by a chatbot powered by ChatGPT, which gives them access to resources and information that would otherwise be hard to come by.

6.13 INDIAN FARMERS WHO MAY NOT BE ACCUSTOMED TO USING A SMARTPHONE MIGHT BENEFIT FROM THE CHATGPT-POWERED WHATSAPP CHATBOT

ChatGPT will be integrated with WhatsApp by the Indian Ministry of Electronics and IT to help farmers in India better grasp government initiatives.

For those who are not comfortable with smartphone typing, the ChatGPT-powered WhatsApp chatbot will respond to questions via voice notes. Government programs will soon be accessible to Indian farmers through a GPT interface, according to a statement made by Microsoft CEO Satya Nadella.

The AI program's present English inputs and lack of support for regional languages, however, may make the integration take longer. To train the AI chatbot, one must build substantial datasets of regional Indian languages.

With the help of OpenAI's Reinforcement Learning from Human Feedback, the ChatGPT platform—which is a component of complex systems that generate content in many forms—was taught.

6.14 CONCLUSION

In order to give farmers and customers information, direction, and help, the agriculture industry is using chatbots—computer systems that simulate human interactions. In addition to providing weather updates, pest control advice, soil test findings, and market pricing information, they can help farmers make quicker decisions about planting, irrigating, harvesting, and applying pesticides. They may also improve supply chain management and logistics for agribusinesses, offer input on preferences, and increase customer trust in agricultural products. In addition, chatbots may provide tailored solutions, environmental insights, and market research.

With chatbots available to answer a variety of agricultural questions, their application in agriculture is growing. AgFunnel, FarmChat [5], ChatGPT, and Krushi [6] – The Farmer Chatbot, are a few examples. An AI chatbot named Krushi helps farmers with seasonal problems, infections, and crop planting. FarmChat is a conversational bot that responds to questions on weather and plant security. AgFunnel is a platform for startup agri-marketing that makes use of chatbots and AI. Engaging talks on a range of topics may be facilitated with ChatGPT.

REFERENCES

1. Usip, Patience U., Edward N. Udo, Daniel E. Asuquo, and Otobong R. James. "A Machine Learning-Based Mobile Chatbot for Crop Farmers." In *International Conference on Electronic Governance with Emerging*

Technologies, eds. Ortiz-Rodríguez, F., Tiwari, S., Sicilia, MA., Nikiforova, A., pp. 192–211. Cham: Springer Nature Switzerland, 2022.

2. Darapaneni, Narayana, Selvakumar Raj, Venkatesh Sivaraman, Sunil Mohan, and Anwesh Reddy Paduri. "LSTM-RASA Based Agri Farm Assistant for Farmers." arXiv preprint arXiv:2204.09717, 2022.
3. Arora, Bhavika, Dheeraj Singh Chaudhary, Mahima Satsangi, Mahima Yadav, Lotika Singh, and Prem Sewak Sudhish. "Agribot: A Natural Language Generative Neural Networks Engine for Agricultural Applications." In *2020 International Conference on Contemporary Computing and Applications* (IC3A), pp. 28–33. IEEE, Lucknow, India, 2020.
4. Kumar, V. V. Sunil, M. V. Thanmayee, K. V. S. Satwik, P. Gayathri, and M. Greeshma. "Agriculture Assistant Chatbot Using CNN." *Journal of Engineering Sciences* 14, no. 05, pp. 60–67, (2023).
5. "FarmChat: Using Chatbots to Answer Farmer Queries in India" www.ictworks.org/farmer-chatbot-india/ January 2, 2019.
6. Momaya, Mihir, Anjnya Khanna, Jessica Sadavarte, and Manoj Sankhe. "Krushi–the Farmer Chatbot." In *2021 International Conference on Communication information and Computing Technology (ICCICT)*, pp. 1–6. IEEE, Mumbai, India, 2021.

Chapter 7

Efficient and Sustainable Farming

Harnessing Drone-Based Intelligent Spraying for Herbicides and Pesticides

Gurpreet Singh, Sukhpreet Singh, and Mohammad Nazmul Alam

7.1 INTRODUCTION

Traditional herbicide and pesticide methods in agriculture suffer from inefficiency, imprecise targeting, and environmental risks. Widespread chemical use leads to overuse, runoff, and harm to non-target organisms [1]. Manual labor is time-consuming and demanding for farmers [2]. Innovative solutions are needed for more effective pest and weed control. Drone-based spraying offers a transformative solution, using advanced tech, AI, and machine learning to precisely target and efficiently apply chemicals [3]. Drones autonomously navigate fields, identify pests, and apply chemicals with high accuracy [4]. This reduces chemical use, and human exposure to hazardous substances, and enhances sustainability [5]. Integrating drones in agriculture optimizes resources, increases yields, and minimizes environmental impact.

7.1.1 Need for Innovation

7.1.1.1 Increasing Demand for Sustainable Agriculture and Reduced Chemical Usage

Global agriculture is shifting toward sustainability. Consumers, regulators, and farmers recognize the need for eco-friendly practices. Traditional spraying methods involve excessive chemical use, causing soil contamination, water pollution, and harm to non-target organisms [1]. Growing public awareness of health risks from pesticide residues in food drives demand for cleaner, safer agriculture [6]. This necessitates innovative approaches promoting sustainable farming and reduced chemical reliance.

DOI: 10.1201/9781003676102-7

7.1.2 Limitations and Drawbacks of Conventional Spraying Techniques

Conventional spraying methods, foundational in pest and weed control, have limitations affecting their effectiveness and sustainability. Indiscriminate chemical use leads to uneven coverage, causing resource wastage and pest resistance [7]. Manual spraying strains workers, impacting their well-being and productivity [2]. Chemical drift due to weather conditions poses risks to nearby ecosystems and communities [8]. These challenges emphasize the need for precise, efficient, and eco-friendly spraying techniques.

7.1.3 Drone-based Intelligent Spraying Systems: An innovative Solution

Drone-based intelligent spraying revolutionizes agriculture, overcoming traditional drawbacks. With advanced tech like remote sensing, AI analytics, and precise navigation, drones make spraying data-driven [3]. Equipped with cameras and sensors, they detect pests and apply chemicals precisely, reducing waste and environmental impact [4]. This optimizes chemical use and enhances pest and weed control efficiency. Drones access challenging terrain, surpassing conventional machinery's limitations [9]. In essence, this approach combines innovation and sustainability, addressing current challenges and promoting responsible and efficient farming. See Figure 7.1 for three UAV types [10].

7.2 PRINCIPLES AND TECHNOLOGY BEHIND DRONE-BASED SPRAYING SYSTEMS

Drone-based intelligent spraying revolutionizes agriculture using advanced technology for precision, efficiency, and sustainability. Unmanned aerial vehicles (drones) equipped with specialized gear and AI-driven software play a central role. It begins with data collection and analysis, using high-resolution aerial imagery from onboard cameras and sensors to pinpoint

(a) (b) (c)

Figure 7.1 Types of UAVs (a) Fixed wing drone (b) Helicopters (c) Multi-copter drone.

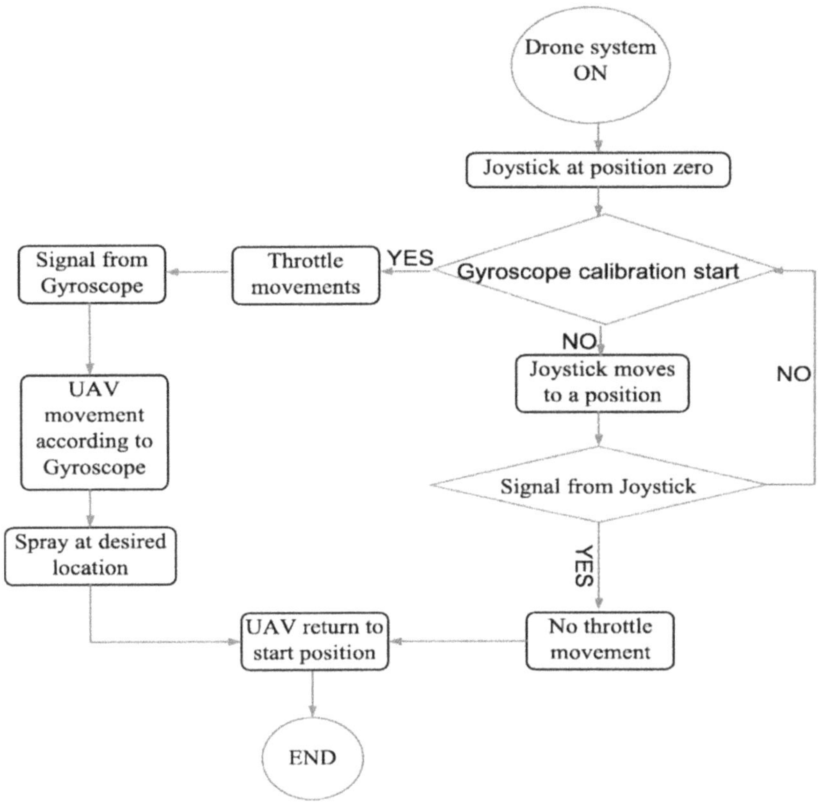

Figure 7.2 Workflow Diagram of Spraying Drone.

treatment areas [11]. The drone's navigation system ensures accuracy and obstacle avoidance. Hovering over the field, the drone's spraying mechanism, typically featuring variable-rate nozzles, dispenses precise herbicides or pesticides only where needed, informed by real-time data analysis [12]. Refer to the spraying drone workflow Figure 7.2 [10].

7.2.1 Integration of Artificial Intelligence and Machine Learning Algorithms for Precise Targeting and Decision-Making

AI and ML algorithms are crucial for the success of drone-based intelligent spraying. AI-driven software processes aerial data, identifying patterns and anomalies related to pest infestations or weed growth [13]. Machine learning algorithms continuously improve their understanding of these patterns, allowing the system to make informed decisions about spray distribution. By learning from past data and adapting to changing conditions, the system optimizes chemical application, preventing over-spraying and reducing environmental impact [14]. This dynamic decision-making ensures each application is customized to the specific needs of a field, effectively addressing variable pest populations and field characteristics.

7.2.3 Benefits of Drones in Terms of Efficiency, Accuracy, and Cost-Effectiveness

Drone-based intelligent spraying transforms agriculture with numerous advantages. Firstly, it significantly boosts efficiency by covering large areas quickly, reducing spraying time compared to traditional methods [3]. This efficiency cuts labor costs and increases productivity. Additionally, the pinpoint accuracy of drones minimizes chemical wastage and non-target exposure [15]. AI and ML algorithms enhance decision-making accuracy, aiding informed chemical application strategies. Reduced human presence in the field enhances operator safety and minimizes chemical exposure [16]. These benefits collectively create a more sustainable and economically viable agricultural ecosystem.

7.3 ADVANTAGES AND APPLICATIONS

7.3.1 Advantages of Using Drone-Based Spraying for Herbicides and Pesticides

Adopting drone-based intelligent spraying revolutionizes agriculture with precision targeting through AI and ML algorithms, reducing chemical usage significantly [17]. This minimizes the risk of chemical runoff, water source contamination, and soil degradation, enhancing environmental sustainability [18]. Drones' pinpoint accuracy reduces chemical wastage, leading to cost savings for farmers [19]. Their speed and efficiency cover larger areas in less time, optimizing labor resources and increasing productivity [20]. Additionally, their ability to navigate challenging terrains enhances versatility [21]. These benefits collectively improve crop protection, resource management, and responsible chemical application. Refer to the Figure 7.3 for a fully automatic pesticide spraying system [10].

7.3.2 Case Studies and Success Stories from Different Agricultural Sectors

Drone-based intelligent spraying proves its efficacy in various agricultural sectors, showcasing its potential to transform modern farming. In crop farming, drones combat pest infestations and optimize herbicide applications

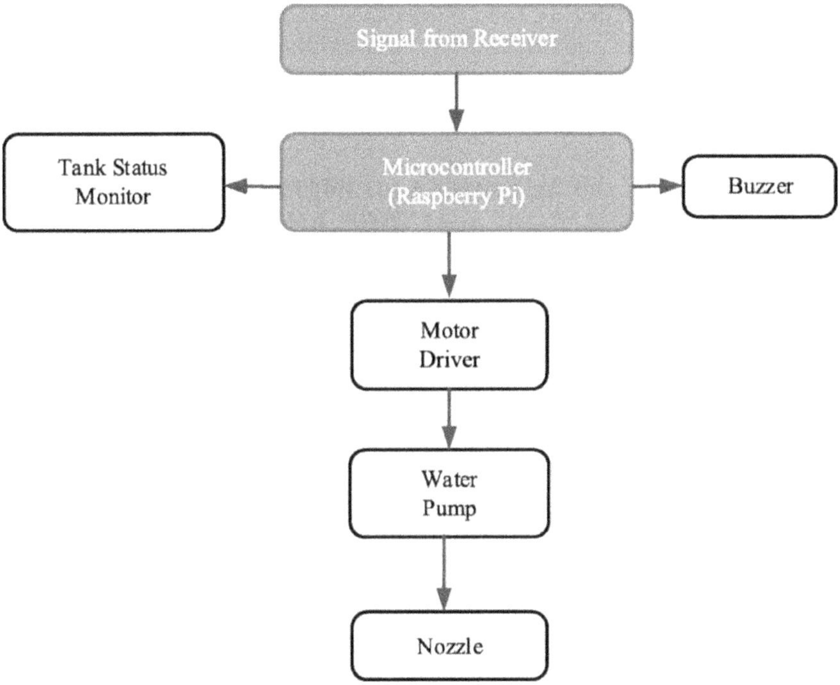

Figure 7.3 Block of a Fully Automatic Pesticide Spraying System.

Figure 7.4 UAV Application in Agricultural Field (a) Field mapping (b) Crop monitoring.

in large fields [22]. In vineyards and orchards, where traditional equipment navigation is challenging, drones provide a precise and adaptable solution for targeted spraying [23]. For example, in vineyards, advanced imaging-equipped drones identify grape clusters needing treatment, reducing over-spray and enhancing accuracy [24]. In orchards, drones efficiently address varying tree heights and foliage densities, ensuring consistent pest control and improved fruit quality [24]. These case studies highlight drone-based spraying's versatility and potential to meet unique requirements in different agricultural contexts. Refer to Figure 7.4 for a UAV application in agricultural field mapping and crop monitoring.

7.4 POTENTIAL IMPACT ON YIELD IMPROVEMENT, REDUCED CHEMICAL WASTAGE, AND ENVIRONMENTAL SUSTAINABILITY

Drone-based intelligent spraying enhances agricultural sustainability and productivity. Accurate delivery of herbicides and pesticides manages pests effectively while minimizing harm to non-target organisms and ecosystems [26]. Reduced chemical wastage yields cost savings and lowers the risk of environmental contamination [27]. Precise targeting improves pest and disease control, leading to healthier crops, higher yields, and better quality [28]. Lower chemical use and minimized drift protect beneficial insects and pollinators, promoting biodiversity and ecological balance [29]. Drone technology aligns with sustainable agriculture, meeting current needs without compromising future generations.

7.5 ADDRESSING CHALLENGES AND CONCERNS

7.5.1 Challenges and Limitations of Drone-Based Spraying

Drone-based intelligent spraying, while promising, presents challenges and limitations that require careful consideration. Regulatory frameworks for drone operations, airspace management, and pesticide application can vary by region, creating legal obstacles [30]. Safety concerns, including potential collisions with manned aircraft or other drones, demand rigorous training and adherence to safety protocols [31]. Public perception matters, since noise pollution, privacy intrusion, and potential ecological impacts can be concerns [32]. Additionally, adverse weather conditions like strong winds or rain can limit the effectiveness and accuracy of drone-based spraying [33]. Addressing these challenges necessitates proactive measures to ensure responsible and effective drone technology use in agriculture.

7.5.2 Strategies for Mitigating Risks and Ensuring Responsible Use

Overcoming the challenges of drone-based spraying requires a multi-pronged approach. Collaboration among agricultural stakeholders, regulatory bodies, and technology providers is crucial to establishing standardized guidelines for safe and responsible drone operations [34]. Training programs for drone operators should cover not only piloting skills but also knowledge of pesticide application, environmental concerns, and compliance with regulatory requirements [35]. The use of advanced sensors and real-time data analysis can enhance a drone's ability to adapt to changing environmental conditions, reducing risks and improving application accuracy [36]. Promoting transparency and clear communication within the agricultural community can help build public trust and understanding of the benefits and safety protocols of drone-based spraying [37].

7.5.3 Concerns Related to Privacy, Data Security, and Non-Target Organisms

The integration of technology in agriculture raises concerns about data security and privacy, as drone operations involve the collection and storage of sensitive information. Farmers and service providers should prioritize data encryption, secure storage, and adherence to data protection regulations [38]. Additionally, the potential impact of pesticide drift and unintended exposure on non-target organisms, such as beneficial insects and wildlife, is a significant concern [39]. By employing advanced imaging systems and AI algorithms, drones can accurately target specific areas, minimizing overspray and reducing the risk of non-target organism exposure [40]. Rigorous testing and validation of drone-based spraying systems can assure their ecological safety and minimize unintended consequences.

7.6 TECHNOLOGICAL INNOVATIONS AND FUTURE TRENDS

7.6.1 Emerging Technologies and Advancements in Drone-Based Intelligent Spraying

The field of drone-based intelligent spraying is advancing rapidly, driven by technological innovations. Precision agriculture benefits from LiDAR (Light Detection and Ranging) sensors that provide high-resolution three-dimensional crop canopy data. Drones with LiDAR can create detailed maps of crop height and structure, improving spray accuracy based on crop growth stages [41]. The adoption of swarm robotics, where multiple drones work together seamlessly, has the potential to optimize coverage and reduce application time, enhancing efficiency for large-scale operations [42]. The development of hybrid drone systems, combining rotary-wing and fixed-wing capabilities, enables longer flight durations and broader coverage of agricultural landscapes [43]. These emerging technologies open new opportunities for improved precision, scalability, and efficiency in drone-based spraying.

7.6.2 The Integration of Sensor Technologies, Advanced Imaging, and Real-Time Data Analysis

Sensor technologies and advanced imaging play a crucial role in refining the accuracy and efficacy of drone-based spraying. Hyperspectral sensors, capable of detecting a wide range of wavelengths beyond visible light, enable the identification of stressors and disease patterns in crops before they become visible to the naked eye [44]. Coupled with multispectral imaging, drones can create detailed vegetation maps, facilitating precise applications based on the specific needs of different areas within a field [45]. Real-time data analysis powered by artificial intelligence enables on-the-fly decision-making, adjusting spraying parameters based on dynamic environmental conditions [46]. This integration enhances targeting accuracy, minimizes chemical wastage, and optimizes resource allocation for sustainable farming practices.

7.6.3 Future Developments and Their Implications for Sustainable Agriculture

The future of drone-based intelligent spraying promises to transform sustainable agriculture. Autonomous navigation systems, including improved obstacle detection and avoidance algorithms, will enhance the safety and reliability of drone operations [47]. Collaboration between drones and ground-based robots equipped with sensors could provide a holistic approach to field monitoring, disease detection, and targeted intervention [48]. The rise of edge computing and decentralized data processing allows drones to conduct real-time analysis on board, reducing reliance on external processing resources [49]. Moreover, the integration of biological control methods, such as releasing beneficial insects, with drone-based spraying, can lead to a holistic approach that minimizes chemical usage while ensuring effective pest management [50, 51]. These innovations have the potential to revolutionize agricultural practices, promoting sustainability, resource efficiency, and improved crop yields.

7.6.3.1 Economic and Environmental Impacts

Drone-based spraying's economic viability relies on initial costs, operational savings, and Return On Investment (ROI). Upfront expenses for drones, sensors, and software may be significant but are offset by long-term benefits. Farmers save on reduced chemical use, increased yields, and decreased labor. Precise targeting minimizes waste. A cost-benefit analysis considering crop type, field size, and conditions is essential for feasibility.

Drone-based spraying revolutionizes farming, offering cost savings and efficiency [52]. Manual spraying is labor-intensive, while drones cover more area in less time. Precise targeting reduces excessive chemical use, cutting costs. Streamlined processes enhance productivity and profitability. Drone-based spraying significantly reduces chemical usage, benefiting the environment [53]. Traditional methods over apply chemicals, causing contamination and pest resistance. Drones, with sensors and real-time data, target treatment areas, reducing environmental impact. Reduced drift and site-specific strategies enhance resource utilization and ecological effects.

The combined economic and environmental benefits make drone-based spraying key to a sustainable and efficient farming future. Reduced chemical waste, improved labor efficiency, and better environmental stewardship are transforming agriculture.

7.7 ADOPTION STRATEGIES AND IMPLEMENTATION

7.7.1 Adoption Process for Farmers and Agricultural Stakeholders

Transitioning to drone-based intelligent spraying requires a well-defined adoption process to ensure its successful integration into existing agricultural practices. Farmers and stakeholders should follow these steps:

1. Assessment: Evaluate the specific needs and challenges of the farm, considering factors such as crop type, field size, topography, and operational requirements [54].
2. Education: Gain a comprehensive understanding of drone technology, its benefits, and potential limitations. Seek information through workshops, seminars, and online resources [55].
3. Needs analysis: Identify the key areas where drone-based spraying can provide the most value, such as precise targeting of herbicides, pest management, and disease control.
4. Equipment selection: Choose the appropriate drone model, sensors, and software that align with the farm's requirements and budget [56].
5. Training: Ensure that drone operators are trained to handle the equipment safely and efficiently, including flight operations, maintenance, and data analysis [57].
6. Implementation plan: Develop a detailed plan outlining how drones will be integrated into existing operations. Define roles and responsibilities, flight schedules, and data management procedures [58].
7. Testing and Adjustment: Pilot the drone-based spraying on a smaller scale to fine-tune operations, identify challenges, and make necessary adjustments.
8. Scaling Up: Gradually expand the use of drone-based spraying as operators gain experience and confidence, potentially covering larger portions of the farm.

7.7.2 Training and Education Programs for Drone Operators and Technicians

Successful adoption of drone-based intelligent spraying hinges on the competence of the operators and technicians. To ensure effective implementation:

1. Basic training: Provide training on drone operations, flight controls, safety protocols, and emergency procedures.
2. Sensors and data analysis: Train operators to interpret sensor data and make informed decisions based on real-time information [59].
3. Maintenance: Offer guidance on equipment maintenance, troubleshooting, and repairs to minimize downtime [60].
4. Regulatory compliance: Familiarize operators with aviation regulations, local laws, and airspace restrictions [61].
5. Continuous Learning: Encourage continuous education through workshops, online courses, and industry events to stay updated on technological advancements and best practices [62].

7.7.3 Strategies for Regulatory Compliance and Collaboration with Governmental Bodies

Regulatory compliance is essential for the responsible and legal use of drone-based spraying systems. Strategies for collaboration with governmental bodies include:

Licensing and Certification: Ensure that drone operators are licensed and certified according to aviation regulations in the region [63].

Airspace Authorization: Obtain necessary permits to operate drones within specified airspace, adhering to height restrictions and no-fly zones.

Environmental Regulations: Understand and comply with regulations related to chemical application, environmental protection, and pesticide use [64].

Data Privacy: Implement data protection measures to safeguard sensitive information collected during drone operations.

Collaboration: Engage with local agricultural departments, aviation authorities, and environmental agencies to foster collaboration and receive guidance on compliance [65].

By following these adoption strategies and collaborating with relevant stakeholders, farmers and agricultural practitioners can seamlessly integrate drone-based intelligent spraying into their operations, enhancing efficiency, sustainability, and productivity.

7.8 CONCLUSION

Drone-based intelligent spraying is a transformative advance in modern agriculture, offering a precise and efficient solution to challenges posed by traditional herbicide and pesticide application. It has the potential to revolutionize agriculture toward greater efficiency and sustainability.

This technology signifies a shift from conventional methods, providing precision, accuracy, and efficiency in herbicide and pesticide application. It utilizes aerial platforms, sensors, and advanced algorithms for targeted delivery, reducing waste, and environmental impact, and optimizing resource use. Adopting drone-based spraying brings benefits like increased productivity, reduced labor, and cost savings. However, it also presents challenges such as regulatory compliance, safety, and public perception that need careful management. Balancing these aspects is crucial for maximizing the technology's potential.

While drone-based intelligent spraying shows promise, its evolution is ongoing. Continued research, development, and collaboration among researchers, industry experts, agricultural organizations, and regulatory bodies are essential to foster responsible innovation. Sharing insights and best practices can collectively shape the technology's path toward sustainable agriculture. Embracing drone-based intelligent spraying with cautious optimism paves the way for a future where precise, efficient, and environmentally conscious herbicide and pesticide application becomes the norm, enhancing agricultural resilience and productivity.

REFERENCES

1. Wauchope, R. D. (1978). The pesticide content of surface water draining from agricultural fields—A review. *Journal of Environmental Quality*, 7(4), 459–472. https://doi.org/10.2134/jeq1978.00472425000700040001x
2. Daberkow, S. G., & McBride, W. D. (2003). Farm and operator characteristics affecting the awareness and adoption of precision agriculture technologies in the US. *Precision Agriculture*, 4(2), 163–177. https://doi.org/10.1023/A:1024557205871
3. Torres-Sánchez, J., López-Granados, F., & Peña, J. M. (2015). An automatic object-based method for optimal thresholding in UAV images: Application for vegetation detection in herbaceous crops. *Computers and Electronics in Agriculture*, 114, 43–52. https://doi.org/10.1016/j.compag.2015.03.019
4. Kokate, P., Middey, A., Sadistap, S., Sarode, G., & Narayan, A. (2024). Review on drone-assisted air-quality monitoring systems. *Drones and Autonomous Vehicles*, 1(1), 10005. https://doi.org/10.35534/dav.2023.10005
5. Beckie, H. J., Blackshaw, R. E., Low, R., Hall, L. M., Sauder, C. A., Martin, S., ... Shirriff, S. W. (2013). Glyphosate- and Acetolactate Synthase Inhibitor–Resistant Kochia (Kochia scoparia) in Western Canada. *Weed Science*, 61(2), 310–318. doi:10.1614/WS-D-12-00140.1

6. Thayer, K. A., Heindel, J. J., Bucher, J. R., Gallo, M. A., & Portier, C. J. (2012). Role of environmental chemicals in diabetes and obesity: A National Toxicology Program workshop review. *Environmental Health Perspectives,* 120(6), 779–789.
7. Somasundaram, L., & Coats, J. R. (Eds.). (1991). Pesticide transformation products. *Fate and significance in the environment (*No. 459, pp. xii+-305).
8. Majewski, M.S. (1996). Pesticides in the Atmosphere: Distribution, Trends, and Governing Factors (1st ed.). CRC Press. https://doi.org/10.1201/9780429062780
9. Sun, Z., Wang, X., Wang, Z., Yang, L., Xie, Y., & Huang, Y. (2021). UAVs as remote sensing platforms in plant ecology: Review of applications and challenges. *Journal of Plant Ecology*, 14(6), 1003–1023. https://doi.org/10.1093/jpe/rtab089
10. Hafeez, A., Husain, M. A., Singh, S. P., Chauhan, A., Khan, M. T., Kumar, N., Chauhan, A., & Soni, S. K. (2023). Implementation of drone technology for farm monitoring & pesticide spraying: A review. *Information Processing in Agriculture*, 10(2), 192–203. https://doi.org/10.1016/j.inpa.2022.02.002
11. Navarro, E., Costa, N., & Pereira, A. (2020). A Systematic Review of IoT Solutions for Smart Farming. *Sensors,* 20(15), 4231. https://doi.org/10.3390/s20154231
12. Wang, J., Wang, Y., Li, G., & Qi, Z. (2024). Integration of Remote Sensing and Machine Learning for Precision Agriculture: A Comprehensive Perspective on Applications. *Agronomy,* 14(9), 1975. https://doi.org/10.3390/agronomy14091975
13. Patel, P. P., & Vaghela, D. B. (2019). Crop diseases and pests detection using convolutional neural network. In *2019 IEEE International Conference on Electrical, Computer and Communication Technologies (ICECCT)* (pp. 1–4). IEEE. https://doi.org/10.1109/ICECCT.2019.8869510
14. Garre, P., & Harish, A. (2018, December). Autonomous agricultural pesticide spraying UAV. *IOP Conference Series: Materials Science and Engineering*, 455(1), 012030. IOP Publishing. https://doi.org/10.1088/1757-899X/455/1/012030
15. Cederlund, H. (2017). Effects of spray drift of glyphosate on nontarget terrestrial plants—A critical review. *Environmental Toxicology and Chemistry*, 36(11), 2879–2886. https://doi.org/10.1002/etc.3925
16. Jasim, A. N., Fourati, L. C., & Albahri, O. S. (2023). Evaluation of unmanned aerial vehicles for precision agriculture based on integrated fuzzy decision-making approach. *IEEE Access,* 11, 75037–75062. https://doi.org/10.1109/ACCESS.2023.3294094
17. Guebsi, R., Mami, S., & Chokmani, K. (2024). Drones in Precision Agriculture: A Comprehensive Review of Applications, Technologies, and Challenges. *Drones,* 8(11), 686. https://doi.org/10.3390/drones8110686
18. Ukaegbu, U. F., Tartibu, L. K., Okwu, M. O., & Olayode, I. O. (2021). Development of a Light-Weight Unmanned Aerial Vehicle for Precision Agriculture. *Sensors,* 21(13), 4417. https://doi.org/10.3390/s21134417

19. Romero, A. M. (2018). Chemical lands: Pesticides, aerial spraying, and health in North America's grasslands since 1945 by David D. Vail. *Environmental History*, 23(4), 920–922. https://doi.org/10.1093/envhis/emy062
20. Chen, H., Lan, Y., Fritz, B. K., Hoffmann, W. C., & Liu, S. (2021). Review of agricultural spraying technologies for plant protection using unmanned aerial vehicle (UAV). *International Journal of Agricultural and Biological Engineering*, 14(1), 38–49. https://doi.org/10.25165/j.ijabe.20211401.5714
21. Singh, R. K., Berkvens, R., & Weyn, M. (2021). AgriFusion: An architecture for IoT and emerging technologies based on a precision agriculture survey. *IEEE Access*, 9, 136253–136283. https://doi.org/10.1109/ACCESS.2021.3116814
22. Li, L., He, X., Song, J., Wang, X., Jia, X., & Liu, C. (2017). Design and experiment of automatic profiling orchard sprayer based on variable air volume and flow rate. *Transactions of the Chinese Society of Agricultural Engineering*, 33(1), 70-76.
23. Barbosa Júnior, M. R., Santos, R. G. d., Sales, L. d. A., & Oliveira, L. P. d. (2024). Advancements in Agricultural Ground Robots for Specialty Crops: An Overview of Innovations, Challenges, and Prospects. *Plants*, 13(23), 3372. https://doi.org/10.3390/plants13233372
24. Gautam, D., Ostendorf, B., & Pagay, V. (2021). Estimation of Grapevine Crop Coefficient Using a Multispectral Camera on an Unmanned Aerial Vehicle. *Remote Sensing,* 13(13), 2639. https://doi.org/10.3390/rs13132639
25. Tahir, M. N., Lan, Y., Zhang, Y., Wenjiang, H., Wang, Y., & Naqvi, S. M. Z. A. (2023). Application of unmanned aerial vehicles in precision agriculture. *Precision Agriculture*, 55–70. https://doi.org/10.1016/B978-0-443-18953-1.00001-5
26. Hamid, S., Jancso, D., Pinter, P., & Takács-Györe, I. (2019). Artificial intelligence for precision agriculture in crop protection: A review. *Computers and Electronics in Agriculture,* 162, 133–146.
27. Shahi, T. B., Xu, C.-Y., Neupane, A., & Guo, W. (2023). Recent Advances in Crop Disease Detection Using UAV and Deep Learning Techniques. *Remote Sensing*, 15(9), 2450. https://doi.org/10.3390/rs15092450
28. Koc, C., Duran, H., & Gerdan Koc, D. (2023). Orchard sprayer design for precision pesticide application. *Erwerbs-Obstbau,* 65(5), 1819–1828. https://doi.org/10.1007/s10341-023-00876-x
29. Crall, J. D., Switzer, C. M., Oppenheimer, R. L., Ford Versypt, A. N., Dey, B., Brown, A., Eyster, M., Guérin, C., Pierce, N. E., Combes, S. A., & de Bivort, B. L. (2018). Neonicotinoid exposure disrupts bumblebee nest behavior, social networks, and thermoregulation. *Science,* 362(6415), 683–686. https://doi.org/10.1126/science.aat1598
30. Spalevic, Z., Ilic, M., & Savija, V. (2018). The use of drones in agriculture: ICT policy, legal and economical aspects. *Ekonomika: Journal for Economic Theory and Practice and Social Issues*, 64(4), 93–107. https://doi.org/10.22004/ag.econ.290358

31. Emimi , M., Khaleel, M., & Alkrash , A. (2023). The Current Opportunities and Challenges in Drone Technology . *International Journal of Electrical Engineering and Sustainability,* 1(3), 74–89. Retrieved from https://ijees.org/index.php/ijees/article/view/47
32. Velusamy, P., Rajendran, S., Mahendran, R. K., Naseer, S., Shafiq, M., & Choi, J.-G. (2022). Unmanned Aerial Vehicles (UAV) in Precision Agriculture: Applications and Challenges. *Energies*, 15(1), 217. https://doi.org/10.3390/en15010217
33. Yu, Y., Lan, Y., Wang, G., Hussain, M., Wang, H., Yu, X., Shan, C., Wang, B., & Song, C. (2023). Evaluation of the deposition and distribution of spray droplets in citrus orchards by plant protection drones. *Frontiers in Plant Science,* 14, Article 1303669. https://doi.org/10.3389/fpls.2023.1303669
34. Rauhala, A., Tuomela, A., & Leviäkangas, P. (2023). Chapter 12 – An overview of unmanned aircraft systems (UAS) governance and regulatory frameworks in the European Union (EU). In D. Bochtis, A. C. Tagarakis, & D. Kateris (Eds.), Unmanned aerial systems in agriculture (pp. 269–285). Academic Press. https://doi.org/10.1016/B978-0-323-91940-1.00012-8
35. Yaqot, M., Menezes, B. C., & Al-Ansari, T. (2021). Unmanned aerial vehicles in precision agriculture towards circular economy: A process system engineering (PSE) assessment. In M. Türkay & R. Gani (Eds.), Computer aided Chemical Engineering (Vol. 50, pp. 1559–1565). Elsevier. https://doi.org/10.1016/B978-0-323-88506-5.50241-2
36. Smyczyński, P., Starzec, Ł., & Granosik, G. (2017). Autonomous drone control system for object tracking: Flexible system design with implementation example. In 2017 22nd International Conference on Methods and Models in Automation and Robotics (MMAR) (pp. 734–738). *IEEE.* https://doi.org/10.1109/MMAR.2017.8046919
37. Antwi, S. H., Stephens, C. G., Rolston, A., Getty, D., & Linnane, S. (2025). Public participation in environmental decision-making: A water sector perspective. *Environmental and Sustainability Indicators*, 26, 100656. Elsevier. https://doi.org/10.1016/j.indic.2025.100656
38. Khan, M. A., Kumar, N., Alsamhi, S. H., Barb, G., Zywiołek, J., Ullah, I., Noor, F., Shah, J. A., & Almuhaideb, A. M. (2025). Security and privacy issues and solutions for UAVs in B5G networks: A review. *IEEE Transactions on Network and Service Management*, 22(1), 892–912. https://doi.org/10.1109/TNSM.2024.3487265
39. Kammerer, M. A., Biddinger, D. J., Rajotte, E. G., & Mortensen, D. A. (2015). Local plant diversity across multiple habitats supports a diverse wild bee community in Pennsylvania apple orchards. *Environmental Entomology,* 45(1), 32–38. https://doi.org/10.1093/ee/nvv147
40. Li, S., Li, J., Yu, S., Wang, P., Liu, H., & Yang, X. (2023). Anti-Drift Technology Progress of Plant Protection Applied to Orchards: A Review. *Agronomy,* 13(11), 2679. https://doi.org/10.3390/agronomy13112679
41. Mishra, P. K., & Rai, A. (2021). Role of unmanned aerial systems for natural resource management. *Journal of the Indian Society of Remote Sensing*, 49(3), 671–679. https://doi.org/10.1007/s12524-020-01230-4

42. Tang, J., Duan, H., & Lao, S. (2023). Swarm intelligence algorithms for multiple unmanned aerial vehicles collaboration: A comprehensive review. *Artificial Intelligence Review*, 56(5), 4295–4327. https://doi.org/10.1007/s10462-022-10281-7
43. Kim, J., Kim, S., Ju, C., & Son, H. I. (2019). Unmanned aerial vehicles in agriculture: A review of perspective of platform, control, and applications. *IEEE Access*, 7, 105100–105115. https://doi.org/10.1109/ACCESS.2019.2932119
44. Gitelson, A. A., Viña, A., Ciganda, V., Rundquist, D. C., & Arkebauer, T. J. (2005). Remote estimation of canopy chlorophyll content in crops. *Geophysical Research Letters*, 32(8). https://doi.org/10.1029/2005GL022688
45. Zhou, X.-G., Zhang, D., & Lin, F. (2021). *UAV Remote Sensing:* An Innovative Tool for Detection and Management of Rice Diseases. IntechOpen. doi: 10.5772/intechopen.95535
46. Machine Learning for Precision Agriculture Using Imagery from Unmanned Aerial Vehicles (UAVs): A Survey. *Drones*, 7(6), 382. https://doi.org/10.3390/drones7060382
47. Rezaee, M. R., Hamid, N. A. W. A., Hussin, M., & Zukarnain, Z. A. (2024). Comprehensive review of drones collision avoidance schemes: Challenges and open issues. *IEEE Transactions on Intelligent Transportation Systems*, 25(7), 6397–6426. https://doi.org/10.1109/TITS.2024.3375893
48. Zhang, C., & Kovacs, J. M. (2012). The application of small unmanned aerial systems for precision agriculture: A review. *Precision Agriculture*, 13(6), 693–712. https://doi.org/10.1007/s11119-012-9274-5
49. Cui, H. (2022). Research on agricultural supply chain architecture based on edge computing and efficiency optimization. *IEEE Access*, 10, 4896–4906. https://doi.org/10.1109/ACCESS.2021.3113723
50. Petro, J. B., Plasse, T. R., & McNulty, J. A. (2003). Biotechnology: Impact on biological warfare and biodefense. *Biosecurity and Bioterrorism: Biodefense Strategy, Practice, and Science*, 1(3), 161–168. https://doi.org/10.1089/153871303769201815
51. Oerke, E. C., Dehne, H. W., Schönbeck, F., & Weber, A. (2012). *Crop Production and Crop Protection: Estimated Losses in Major Food and Cash Crops*. Elsevier.
52. Maes, W. H., & Steppe, K. (2019). Perspectives for remote sensing with unmanned aerial vehicles in precision agriculture. *Trends in Plant Science*, 24(2), 152–164. https://doi.org/10.1016/j.tplants.2018.11.007
53. Inoue, Y. (2020). Satellite- and drone-based remote sensing of crops and soils for smart farming – a review. *Soil Science and Plant Nutrition*, 66(6), 798–810. https://doi.org/10.1080/00380768.2020.1738899
54. Daponte, P., De Vito, L., Glielmo, L., Iannelli, L., Liuzza, D., Picariello, F., & Silano, G. (2019). A review on the use of drones for precision agriculture. *IOP Conference Series: Earth and Environmental Science*, 275(1), 012022. https://doi.org/10.1088/1755-1315/275/1/012022
55. Torres-Sánchez, J., López-Granados, F., & Peña, J. M. (2015). An automatic object-based method for optimal thresholding in UAV images: Application

for vegetation detection in herbaceous crops. *Computers and Electronics in Agriculture,* 114, 43–52. https://doi.org/10.1016/j.compag.2015.03.019
56. Tahir, M. N., Lan, Y., Zhang, Y., Wenjiang, H., Wang, Y., Naqvi, S. M. Z. A., & Zaman, Q. (2023). Chapter 4 - Application of unmanned aerial vehicles in precision agriculture. In *Precision Agriculture* (pp. 55–70). Academic Press. https://doi.org/10.1016/B978-0-443-18953-1.00001-5
57. O'Dea, N., Araújo, M. B., & Whittaker, R. J. (2006). How well do Important Bird Areas represent species and minimize conservation conflict in the tropical Andes? *Diversity and Distributions*, 12(2), 205–214. https://doi.org/10.1111/j.1366-9516.2006.00235.x
58. Sampedro, C., Bavle, H., Rodríguez-Ramos, A., Carrio, A., Suárez Fernández, R. A., Sanchez-Lopez, J. L., & Campoy, P. (2017). A fully-autonomous aerial robotic solution for the 2016 International Micro Air Vehicle competition. In *2017 International Conference on Unmanned Aircraft Systems (ICUAS)* (pp. 989–998). IEEE. https://doi.org/10.1109/ICUAS.2017.7991442
59. Mogili, U. M. R., & Deepak, B. B. V. L. (2018). Review on application of drone systems in precision agriculture. *Procedia Computer Science,* 133, 502–509. https://doi.org/10.1016/j.procs.2018.07.063
60. Mirri, S., Prandi, C., & Salomoni, P. (2019). Human-drone interaction: State of the art, open issues and challenges. In *Proceedings of the ACM SIGCOMM 2019 Workshop on Mobile AirGround Edge Computing, Systems, Networks, and Applications* (MAGESys'19, pp. 43–48). Association for Computing Machinery. https://doi.org/10.1145/3341568.3342111
61. Telli, K., Kraa, O., Himeur, Y., Ouamane, A., Boumehraz, M., Atalla, S., & Mansoor, W. (2023). A Comprehensive Review of Recent Research Trends on Unmanned Aerial Vehicles (UAVs). *Systems*, 11(8), 400. https://doi.org/10.3390/systems11080400
62. Rao, S., & Neethirajan, S. (2025). Computational Architectures for Precision Dairy Nutrition Digital Twins: A Technical Review and Implementation Framework. *Sensors*, 25(16), 4899. https://doi.org/10.3390/s25164899
63. López-López, P. C., Oñate, P., & Rocha, Á. (2020). Social media mining, debate and feelings: Digital public opinion's reaction in five presidential elections in Latin America. *Cluster Computing*, 23(3), 1875–1886. https://doi.org/10.1007/s10586-020-03072-8
64. Sishodia, R. P., Ray, R. L., & Singh, S. K. (2020). Applications of Remote Sensing in Precision Agriculture: A Review. *Remote Sensing,* 12(19), 3136. https://doi.org/10.3390/rs12193136
65. Patrício, D. I., & Rieder, R. (2018). Computer vision and artificial intelligence in precision agriculture for grain crops: A systematic review. *Computers and Electronics in Agriculture*, 153, 69–81. https://doi.org/10.1016/j.compag.2018.08.001

Chapter 8

IoT-Based Smart Irrigation Systems

Augmenting Water Management for Crops

Swati Patel, Satyen Parikh, Savan Patel, and Haresh Patel

8.1 INTRODUCTION

Agriculture plays a vital role in a country's economic foundation. The agriculture sector can be considered the backbone of India's economy as 70% of rural households depend on it. According to DAFW (Department of Agriculture and Farmer Welfare), agriculture shares about 18.8% of Indian GVA (Gross Value Added) in 2021–22. Among the total population, 58% of employment is provided by the agriculture sector.

Figure 8.1 shows the percentage share of GVA by the agriculture sector in total GVA for the last 10 years (2012–13 to 2021–22). As shown in the figure the highest GDP covered by the agriculture sector is during 2020–21. This means that during the Corona pandemic it plays a crucial role in the Indian economy [1]. The use of modern technologies in the agriculture sector can increase the shares of agriculture in the Indian economy. The rapid increase in world population also increases food consumption. The changing consumer demands also become challenging for agriculture industries to develop techniques and practices, which satisfy the increasing needs and demands [2]. The very crucial parameter of agricultural production can be considered to be irrigation. The amount of water supplied to crops, the time interval between irrigations, and the crop water needed according to soil conditions all affect crop production. The very challenging issue the agriculture sector is facing nowadays is water scarcity. Figure 8.2 shows the agricultural water sources in India [3].

The major sources can be considered to be groundwater with 46% of agricultural land irrigated by Tubewells and 16 % by other kinds of wells.

Technical advancement in the agriculture sector especially in irrigation can be helpful to increase crop production. The available Smart Irrigation Systems automate the irrigation systems and supply the right amount of water to the crop. The IoT-based SIS uses different sensor data as well as weather data to schedule the irrigation. The stored data can be used to predict future irrigation needs with the help of AI [4–5]. The IoT can be thought of as an extension of the current internet to all devices that can

DOI: 10.1201/9781003676102-8

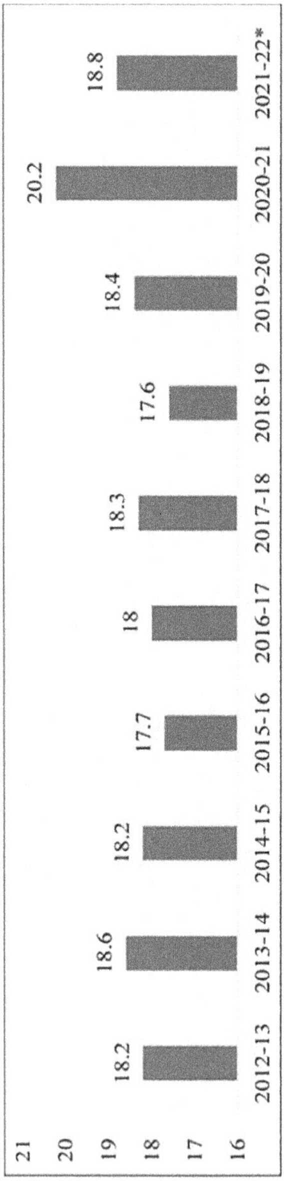

Figure 8.1 Percentage Share of Agriculture and Allied Sector in Total GVA (Source: DAFW.).

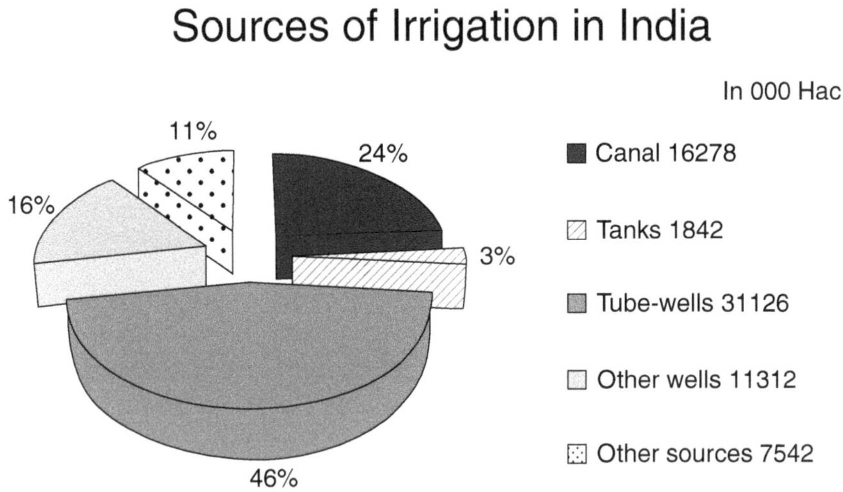

Figure 8.2 Sources of Irrigation in India.

communicate with electronic equipment and are linked to the internet, making devices user-friendly and easy to manage. The IoT is connected to automate the different sections of farming and agricultural processes to make the intact system more fruitful and competent [6]. The revolution in the agricultural sector also eases the farmer's life.

8.2 IMPORTANT PARAMETERS FOR IRRIGATION

Identifying key parameters of any system is a crucial task. Observing key parameters can become efficient and impactful for designing and implementing a new system. The key parameters for any SIS are considered to be soil type, weather conditions, crop growth stage, and irrigation management [7]. As a first step of the observation process of key parameters, we collect data from different sensors, weather parameters, plant growth status, and so forth. The use of the IoT becomes effective and efficient for the observation process. All the collected information is transmitted for further processing [8].

The key parameters for SIS are explained below.

8.2.1 Soil Type

Soil moisture is a very important factor for crop growth. Observing and controlling the moisture level of soil leads to crop productivity and profit gain for farmers. The soil moisture sensors are deployed at the root zone of the crop to read the moisture level of the soil precisely. The data read by the soil moisture sensors are further forwarded to the controller[9].

The automatic irrigation system works in two different ways: One works for scheduling irrigation and the other is just an auto ON/OFF watering system. The scheduling of irrigation required programming according to the system's needs. The auto ON/OFF system is an auto cut-off system that works based on a predefined threshold value [10].

The type of soil also affects irrigation management. For example, sandy soil needs frequent or more irrigation compared to loamy soil.

8.2.2 Weather Conditions

To make the irrigation scheduling more accurate the weather data is also collected before scheduling the irrigation. Some automated systems schedule the irrigation based on weather parameters only. These types of systems are used mainly for turf grass irrigation. The weather-based controllers are also available for use to irrigate the grassland. The controllers work based on a pre-defined schedule or based on auto cut-off mode [11].

Different weather sensors collect weather data like humidity, temperature, and wind speed. The weather applications are also used to collect the weather data of the farm area or grassland. The weather-based smart irrigation system takes the irrigation decision as per the values of ETo (Evapotranspiration).

8.2.3 Crop Growth Stages

The crop growth stage is an important factor for deciding the irrigation scheduling. Different crops have different growth periods based on the stage of growth and have different watering requirements. The consideration of irrigation according to the growth stage increases the crop yield and makes an efficient irrigation system. The Crop Growth stage base irrigation system uses K_c (Crop Coefficient) for deciding the irrigation. K_c is calculated from various crop growth stages [12]. Refer to Figure 8.3 to understand the crop growth stages.

K_c remains constant during the initial stage and mid-growth stage. K_c varies linearly concerning the previous stage K_c during the crop development stage and crop end stage.

$$K_{ci} = K_{cprev} + \left[\frac{i - \Sigma\left(L_{prev}\right)}{L_{stage}}\right]\left(K_{cnext} - K_{cprev}\right)_{(FAO)}$$

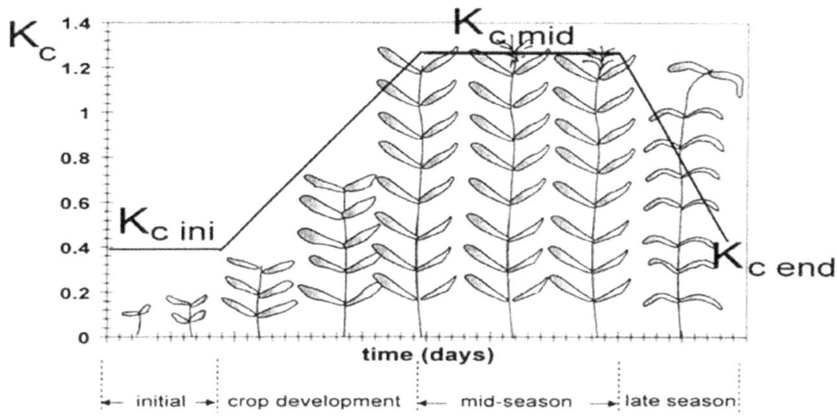

Figure 8.3 Crop Growth Stages.

Where,

> i is the day number within the growing season [taking values from 1 up to the length of the growing season],
> Kci is the crop coefficient on day i,
> L_{stage} is the length of the stage under consideration [days],
> (L_{prev}) the sum of the lengths of all previous stages [days].

The above equation is applied to calculate K_c. The K_c values for various crops were calculated by FOA (Food and Agriculture Organization of the United Nations) [13].

8.2.4 Water Management

Water management in the agricultural sector needs attention as water scarcity is a booming issue for all agricultural nations. As data shown in Figure 8.2, proper agricultural water management can save the use of groundwater. The application of the right amount of water leads to higher productivity [14]. To determine optimum-level irrigation efficiency, the economist would want to know the value of irrigation water and the cost of increased control or management that would permit a reduction in diversion. As water becomes scarce, increasing crop water productivity or reducing diversions would make sense if the water 'saved' could be put to higher-valued uses. [15].

Figure 8.4 represents the projected value of combined water usage by different sectors. The irrigation usage remains high all the time. It depicts

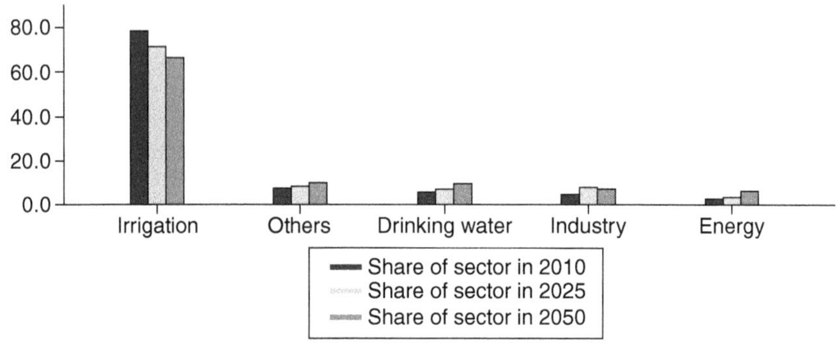

Figure 8.4 Irrigation Water Projection (Source: CWC, NCIWS).

that the proper water management for crop irrigation helps to save ground water [16].

8.3 IOT AND SMART SYSTEMS USED IN IRRIGATION

The recent advances in sensors and evolution in WSN (Wireless Sensor Networks) makes the use of the IoT in agriculture systems very popular.

The outcome of technology in farming is well explained with automated or smart irrigation systems. These systems transform the way of distributing efficient water to crops and higher quality and quantity in yields. Recent irrigation methods supply water without wasting resources. This evolution blesses the farmers in the area where water scarcity is a booming issue in agriculture [17, 18].

The current available smart irrigation systems use different sensors like temperature, soil moisture, humidity, and controllers like the ET controller. All this available technology helps to provide the necessary amount of water to the crop.

The use of IoT is steeply increasing in agriculture. LPWAN is widely used with IoT technologies. As sensor nodes are deployed in open fields, it is a low-power solution for agricultural use. LoRaWAN can become a good service provider in agriculture due to its long range and low power consumption. Scalability is considered one of the irrigation challenges that can be overcome by LoRaWAN. The researchers also show that government support can increase the participation of farmers in technology adoption [19, 20]. The use of various algorithms takes place to predict water supply and to calculate the water needed. The different sensor data such as temperature, humidity, and soil moisture are collected and analyzed for the prediction of water scheduling [21].

8.3.1 Sensor Technology

Different sensors such as temperature, humidity and soil moisture play a vital role in the smart irrigation system.

Temperature and humidity sensor: This sensor is used to derive the surrounding temperature and humidity of farms. This data is used for decision-making on irrigation scheduling.

Soil moisture sensor: The requirement for watering need for the crop can be sensed by sensing the moisture level of the soil and based on this sensing value smart irrigation can be implemented. There are mainly two types of moisture sensors available: Volumetric water content-based sensors and soil tension-based sensors. To increase the accuracy of irrigation scheduling both types of sensor can be used. Water content-based sensors (volumetric sensors) indicate daily water uptake and show the optimum water level for the root. Water tension-based sensors indicate the water-carrying capacity of the soil.

8.3.2 Communication Technology

The IoT–The Internet of Things is the concept of establishing the connection of the devices to the internet or other devices through the internet. The connection between devices and the internet furnishes the communication channel for data and information exchange. IoT devices sense and gather information about their environment, usage patterns, and user inputs, then transmit this information to the cloud. All the data and information gathered in the cloud are analyzed using different AI or ML (Machine Learning) algorithms [22]. The different communication media used for IoT-based irrigation are GSM, LTE, LoRa, and the like.

8.3.3 Cloud Technology

Data collection is a crucial task of SIS. The collected data needs to be processed to make irrigation decisions. Data storage plays an important role at this stage. The database can be on a local machine or in the cloud. Cloud technology provides free as well as paid options for saving, calculating, and presenting data on different devices and platforms [23]. In smart irrigation systems data is collected from different sensors, stored, and processed and users can access the information by connecting with the cloud [24]. The stored cloud data is later accessed for different decision-making for various purposes like extended research and applying AI and ML. The alerts always become helpful to mitigate certain risks and the irrigation scheduling activities become more accurate and efficient [25].

8.3.4 Phases of the Smart Irrigation System

The Smart Irrigation System works in different phases such as data collection, data analysis, processing, and peripheral controlling. Figure 8.5 depicts the phases of SIS.

Data collection: All the required data is collected in this phase. Sensor data, weather application data, crop data, relays, and valve data are collected. The sensors can be soil moisture sensors, temperature, and humidity sensors, and so forth. Data may be collected from weather applications like rain prediction, humidity, temperature, and the like.. The use of sensors and weather applications can vary as per the need of any specific system. No SIS can become functional without customization support as it is deploying in an open environment with great diversity of weather and soil [26].

Data analysis: The collected data is analyzed in relation to some predefined values. The data is collected from different sensors so it should be analyzed to remove unwanted noise and make the data robust for the next stage. The different soils have different wilting points and according to crop and type of soil, the irrigation-starting threshold can be varied.

Processing: This phase processes all the collected data to make an irrigation decision. This phase ensures that the irrigation need for crops is considered and schedules the irrigation. The irrigation need depends on crop type, soil type, and weather conditions. If all the parameter values indicate the deployment of irrigation then the decision is taken to irrigate the relevant crop.

Peripheral controlling: The last phase of the SIS is peripheral controlling. If the processing phase decides to irrigate the crop the next task is controlling peripheral devices associated with this crop irrigation. This phase includes controlling the associated valves and relays to the ON/OFF irrigation cycle.

8.3.5 Basic Architecture

The basic architecture of the Smart Irrigation System is an architecture layout of SIS, which depicts the various component of SIS as shown in Figure 8.6 [27, 28].

- The Arduino Uno microcontroller is used in the system. It contains an ADC (Analog to Digital Converter), a GPIO (General Purpose Input Output), and Communication Ports.

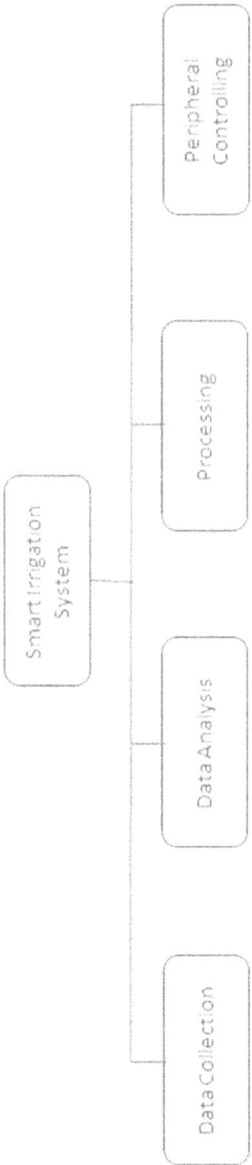

Figure 8.5 Phases of Smart Irrigation System.

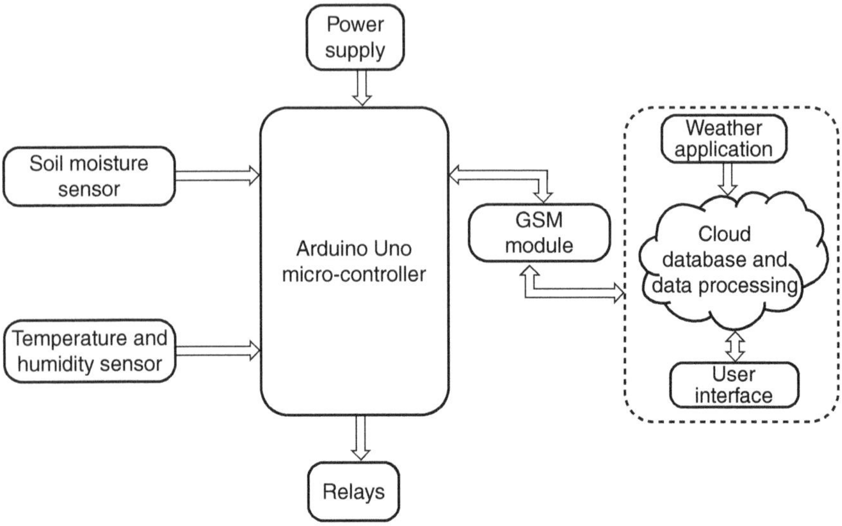

Figure 8.6 Basic Architecture of SIS.

- Soil moisture sensors and temperature and humidity sensors read the data and transfer it to the controller. The data is in analog form so the ADC converts this data to digital form.
- Signal conditioning removes unwanted noise from the analog data.

The controller is supported by battery backup for the power supply.

- The GSM module is used to establish an internet connection between the controller and the cloud.
- Weather data like rain prediction, temperature, and humidity of a particular area are received from weather applications.
- All the received data is stored in a database in the cloud.
- Various algorithms and techniques can be applied to perform processing on stored data to take irrigation decisions.
- The user interface controls the system as well as providing data such as crop type, growth stage, and so forth. It is also used to display notification messages.
- The relay driver receives ON/OFF messages according to the scheduling decision taken by the data processing unit.

8.3.6 Security Concerns

The wireless sensor network security services should ensure the protection of information communicated over the network. The utmost significant security requirements in WSN are considered to be data confidentiality, data integrity, authentication, time synchronization, and the like. [29].

Because there could be a variety of dangers, ensuring security when implementing IoT devices could be challenging. Cloning, insecure software, the leakage of personal data, firmware assaults, routing attacks, denial of service attacks, eavesdropping attacks, and many more security concerns could occur and have negative effects. It has become crucial to create security solutions that might reduce overall difficulties in light of all of these risks. [30].

The research community paid close attention to and addressed several levels of security and privacy concerns in the IoT. Yang et al [31] conducted a four-part survey of the security and privacy issues in the IoT. They begin by highlighting the drawbacks of using security in IoT devices (such as battery life and processing resources) and the suggested fixes (such as a lightweight encryption technique created for embedded systems). Secondly, they list the many IoT attack categories (such as physical, remote, local, and so forth). Thirdly, they concentrate on the processes and architectural frameworks created and used for authentication and authorisation. Finally, they examine security concerns on several layers (such as physical, network, and so forth.).

8.4 IMPACT AND CHALLENGES OF SMART IRRIGATION SYSTEM

8.4.1 Impact of Smart Irrigation System

Technical advancement in any area always comes with a benefit. India is a nation of agriculture, the majority of households, 57.8% engaged in farming activities. The advancement in any agricultural technique always has an impact on the whole society in terms of financial, social and conservational gains [32]. The life of an Indian farmer is filled with many difficulties. Despite their hard work, they often live at the edge of poverty. The farmer is often mentioned as 'The Father of Universe' but they are frequently struggling for basic living needs. Small and marginal farmers live like 'born in debt, live in debt, and die in debt.' Farmers are also less respected in society compared to other professions.

Worldwide, farming is a profession with a higher risk of suicide due to worse conditions than other professions [33]. According to the 2011–12 survey released in 2016 by the Union Ministry of Agriculture & Farmers' Welfare, the average age of farmers in India is 50.1. Currently, youth are not interested in farming activities due to more hard work and lessened profit compared to other professions. According to a report by the World Bank, it is projected that by 2050 the percentage of agricultural workers in the total workforce will crash to 25.7%, which was 58.2% in 2001 [34].

All of these issues can be reduced by technical advancement. Irrigation is considered to be a highly time-consuming activity. Automation in irrigation can become a blessing for older farmers. Women farmers also help to retain the interest of the youth in farming. The most time-consuming task is now a time-saving task with minimal manual intervention.

The use of SIS reduces the annual water bill by preventing water waste. The efficient use of water in crop growth leads to higher productivity. Ultimately, this can increase the farmer's profit and the economy of the nation. It can transform the socio-economic position of farmers.

The smart irrigation system also has conservational benefits such as:

- It reduces the wastage of groundwater capital.
- It helps to reduce soil deprivation.
- It can prevent overwatering, thus keeping plants healthier, making roots stronger, and decreasing weed, disease, and fungus growth. It can thus reduce the use of pesticides.

8.4.2 Challenges of Smart Irrigation System

Challenges in the Adoption of Modern Technology in Agriculture [35, 36]

- Lack of consciousness of technology and missing the proficiency to use.
- Poor infrastructure in terms of storage capacity and transport availability.
- Not practical economically.
- Soil conditions, climatic conditions like improper rainfall, frost, flood, storm, and the like.
- Poor interest of youth in agriculture due to socio-psychological issues.
- Acceptance of new technology by the elderly or illiterate farmers.

8.5 CONCLUSION

Water management is a crucial part of any irrigation system. The use of advanced technology like IoT, sensors, and cloud technology helps augment water management in smart irrigation systems. Researchers are continuously trying to augment irrigation technology and provide the utmost functional efficiency whilst achieving essential performance outcomes. The use of sensors and weather data estimates the real water needs of crops according to their growth stages. Automated irrigation benefited society as it eased farming for elderly farmers and women farmers and also retained the interest of the youth in farming. SIS eliminates the overwatering of crops, which leads to water savings and higher crop productivity. These technologies enable farmers to make informed decisions, conserve resources, and contribute to profitable and sustainable agriculture by leveraging the power of data and automation. The use of crop coefficients enhances the estimation efficiency of irrigation water needs as per the crop growth stage. Adopting new technology in agriculture will ultimately help to make farming more sustainable and productive.

REFERENCES

1. Beckman, J., & Countryman, A. M. (2021). The importance of agriculture in the economy: Impacts from COVID-19. *American Journal of Agricultural Economics*, *103*(5), 1595–1611.
2. Obaideen, K., Yousef, B. A., AlMallahi, M. N., Tan, Y. C., Mahmoud, M., Jaber, H., & Ramadan, M. (2022). An overview of smart irrigation systems using IoT. *Energy Nexus*, Vol 5, 100124.
3. DAFW and MAFW Annual Reports, 2021-22. https://agriwelfare.gov.in/Documents/annual-report-2021-22.pdf
4. Tantalaki, N., Souravlas, S., & Roumeliotis, M. (2019). Data-driven decision making in precision agriculture: The rise of big data in agricultural systems. *Journal of Agricultural & Food Information*, *20*(4), 344–380.
5. Weersink, A., Fraser, E., Pannell, D., Duncan, E., & Rotz, S. (2018). Opportunities and challenges for big data in agricultural and environmental analysis. *Annual Review of Resource Economics*, *10*, 19–37.
6. Nawandar, N. K., & Satpute, V. R. (2019). IoT based low cost and intelligent module for smart irrigation system. *Computers and Electronics in Agriculture*, *162*, 979–990.
7. Patel, S. V., Parikh, S., Patel, S., & Patel, H. (2022, November). Automated irrigation scheduling for different crop using IoT. In *2022 Sixth International Conference on I-SMAC (IoT in Social, Mobile, Analytics and Cloud)(I-SMAC)* (pp. 100–105). IEEE.
8. Garcia Paucar, L., Ramirez Diaz, A., Viani, F., Robol, F., Polo, A., & Massa, A. (2016). *Decision Support for Smart Irrigation by Means of Wireless Distributed Sensors*. Institute of Electrical and Electronics Engineers.
9. Roy, D. K., & Ansari, M. H. (2014). Smart irrigation control system. *International Journal of Environmental Research and Development*, *4*(4), 371–374.
10. Goap, A., Sharma, D., Shukla, A. K., & Krishna, C. R. (2018). An IoT based smart irrigation management system using Machine learning and open source technologies. *Computers and Electronics in Agriculture*, *155*, 41–49.
11. Bwambale, E., Abagale, F. K., & Anornu, G. K. (2022). Smart irrigation monitoring and control strategies for improving water use efficiency in precision agriculture: A review. *Agricultural Water Management*, *260*, 107324.
12. Fereres, E., Goldhamer, D. A., & Parsons, L. R. (2003). Irrigation water management of horticultural crops. *HortScience*, *38*(5), 1036–1042.

13. **Single crop coefficient (K_c) - ET_c Chapter 6** - www.fao.org/3/X0490E/x0490e0b.htm
14. Holzapfel, E. A., Pannunzio, A., Lorite, I., de Oliveira, A. S. S., & Farkas, I. (2014). Design and management of irrigation systems: Chile. *Sustainable Micro Irrigation: Principles and Practices,* 147, pp. 17–25.
15. Kijne, J. W., Barker, R., &Molden, D. (2003). Improving water productivity in agriculture: Editors' overview. *Water Productivity in Agriculture: Limits and Opportunities for Improvement*, xi–xix.
16. Sharma, B. R., Gulati, A., Mohan, G., Manchanda, S., Ray, I., & Amarasinghe, U. (2018). *Water Productivity Mapping of Major Indian Crops*. NABARD.
17. See [2] above.
18. Sudharshan, N., Karthik, A. K., Kiran, J. S., & Geetha, S. (2019). Renewable energy based smart irrigation system. *Procedia Computer Science*, *165*, 615–623.
19. Singh, D. K., Sobti, R., Jain, A., Malik, P. K., & Le, D. N. (2022). LoRa based intelligent soil and weather condition monitoring with internet of things for precision agriculture in smart cities. *IET Communications*, *16*(5), 604–618.
20. Haxhibeqiri, J., De Poorter, E., Moerman, I., & Hoebeke, J. (2018). A survey of LoRaWAN for IoT: From technology to application. *Sensors,* *18*(11), 3995.
21. Reddy, K. S. P., Roopa, Y. M., LN, K. R., &Nandan, N. S. (2020, July). IoT based smart agriculture using machine learning. In *2020 Second International Conference on Inventive Research in Computing Applications (ICIRCA)* (pp. 130–134). IEEE.
22. Qazi, S., Khawaja, B. A., & Farooq, Q. U. (2022). IoT-equipped and AI-enabled next generation smart agriculture: A critical review, current challenges and future trends. *IEEE Access*, *10*, 21219–21235.
23. Salvi, S., Jain, S. F., Sanjay, H. A., Harshita, T. K., Farhana, M., Jain, N., & Suhas, M. V. (2017, February). Cloud based data analysis and monitoring of smart multi-level irrigation system using IoT. In *2017 International Conference on I-SMAC (IoT in Social, Mobile, Analytics and Cloud)(I-SMAC)* (pp. 752–757). IEEE.
24. Tyagi, S., Obaidat, M. S., Tanwar, S., Kumar, N., & Lal, M. (2017, December). Sensor cloud based measurement to management system for precise irrigation. In *GLOBECOM 2017-2017 IEEE Global Communications Conference* (pp. 1–6). IEEE.
25. Sales, N., Remédios, O., & Arsenio, A. (2015, December). Wireless sensor and actuator system for smart irrigation on the cloud. In *2015 IEEE 2nd World Forum on Internet of Things (WF-IoT)* (pp. 693–698). IEEE.
26. Chikankar, P. B., Mehetre, D., & Das, S. (2015, January). An automatic irrigation system using ZigBee in wireless sensor network. In *2015 International Conference on Pervasive Computing (ICPC)* (pp. 1–5). IEEE.
27. Yasin, H. M., Zeebaree, S. R., & Zebari, I. M. (2019, April). Arduino based automatic irrigation system: Monitoring and SMS controlling. In *2019 4th Scientific International Conference Najaf (SICN)* (pp. 109–114). IEEE.

28. Taneja, K., & Bhatia, S. (2017, June). Automatic irrigation system using Arduino UNO. In *2017 International Conference on Intelligent Computing and Control Systems (ICICCS)* (pp. 132–135). IEEE.
29. Sen, J. (2012). Security in wireless sensor networks. *Wireless Sensor Networks: Current Status and Future Trends*, *407*, 408.
30. Rao, R. N., & Sridhar, B. (2018, January). IoT based smart crop-field monitoring and automation irrigation system. In *2018 2nd International Conference on Inventive Systems and Control (ICISC)* (pp. 478–483). IEEE.
31. Yang, Y., Wu, L., Yin, G., Li, L., & Zhao, H. (2017). A survey on security and privacy issues in Internet-of-Things. *IEEE Internet of things Journal*, *4*(5), 1250–1258.
32. Kumar, R., Mishra, R., Gupta, H. P., & Dutta, T. (2021). Smart sensing for agriculture: Applications, advancements, and challenges. *IEEE Consumer Electronics Magazine*, *10*(4), 51–56.
33. Merriott, D. (2016). Factors associated with the farmer suicide crisis in India. *Journal of Epidemiology and Global Health*, 6(4), 217–227.
34. Smelser, N. J., & Baltes, P. B. (Eds.). (2001). *International Encyclopedia of the Social & Behavioral Sciences* (Vol. 11). Elsevier.
35. Katke, K. (2019). Precision agriculture adoption: Challenges of Indian agriculture. *International Journal of Research and Analytical Reviews*, *6*(1), 863–869.
36. Kumar, R., Mishra, R., Gupta, H. P., & Dutta, T. (2021). Smart sensing for agriculture: Applications, advancements, and challenges. *IEEE Consumer Electronics Magazine*, *10*(4), 51–56.

Chapter 9

Robotics in Agriculture to Deal with the Labour Crisis

Saket Joshi

9.1 INTRODUCTION

The boundaries of agriculture are not confined within clear-cut lines. While traditional aspects like soil preparation, planting, cultivation, and harvesting are evident, the post-harvest processes and animal-based activities also intertwine with the broader domain. As technology progresses, the integration of robotics into agriculture emerges as a transformative force, redefining conventional farming practices and addressing challenges such as labor shortages.

9.1.1 Historical Context

The evolution of agriculture is a testament to human ingenuity and adaptability, with a rich history dating back thousands of years. From the earliest forms of subsistence farming to the modern era of advanced technologies, agriculture has undergone profound transformations, often shaped by societal, technological, and environmental shifts. Understanding this historical context provides valuable insights into the challenges and opportunities that have paved the way for the integration of robotics in agriculture.

9.1.1.1 Primitive Beginnings and Agrarian Societies

The origins of agriculture can be traced back to ancient civilizations that transitioned from hunting and gathering to cultivating crops and domesticating animals [1]. The Neolithic revolution marked a turning point in human history, as communities shifted from nomadic lifestyles to settled agrarian societies (Figure 9.1). Innovations such as the plow, irrigation systems, and crop rotation techniques revolutionized farming practices and increased agricultural productivity. Labor-intensive methods prevailed, relying heavily on human and animal power.

DOI: 10.1201/9781003676102-9

Figure 9.1 Agrarian Societies.

9.1.1.2 Industrial Revolution and Mechanization

The 18th and 19th centuries witnessed the Industrial Revolution, which brought about significant advancements in manufacturing and transportation [2]. This era had a profound impact on agriculture as well. The mechanization of farming emerged, introducing tools such as the reaper, thresher, and steam-powered tractors. These innovations reduced the reliance on manual labor and marked the beginning of mechanized agriculture. However, these technologies were limited in their scope and scale.

9.1.1.3 Green Revolution and Technological Leap

The mid-20th century witnessed another transformative phase in agriculture, known as the Green Revolution [3]. This movement introduced high-yielding crop varieties, synthetic fertilizers, and pesticides, leading to remarkable increases in food production. While these advancements addressed pressing issues of food security, they also heightened concerns about environmental sustainability, resource depletion, and labor challenges.

9.1.1.4 The Digital Age and Precision Agriculture

The late 20th century brought forth the digital age, ushering in a new era of technological innovation [4]. Computers, sensors, and data-driven approaches gained prominence in agriculture, leading to the concept of precision agriculture. This approach involved the use of technology to optimize the use of resources, minimize waste, and enhance crop yields. Automation gradually found its way into agricultural practices, with GPS-guided tractors and drones aiding in field management.

9.1.1.5 Integration of Robotics

As we enter the 21st century, the integration of robotics into agriculture represents the latest chapter in its evolution [5]. The convergence of robotics, artificial intelligence, and the Internet of Things has opened new frontiers for the industry. Robotic systems designed to perform tasks ranging from planting and harvesting to data collection and analysis have emerged. This integration addresses challenges such as labor shortages, changing demographics, and the need for sustainable farming practices.

In summary, the historical context of agriculture underscores its resilience and adaptability over the millennia. From primitive agricultural practices to mechanization, technological leaps, and the integration of robotics, the journey of agriculture has been marked by the quest for efficiency, sustainability, and productivity. The current era of robotics in agriculture is a culmination of these efforts, addressing the labor crisis while paving the way for a technologically advanced and sustainable future.

9.1.2 Introduction to Robotics

Robotics is a dynamic interdisciplinary field that encompasses the design, creation, operation, and study of robots. A robot is an autonomous or semi-autonomous machine programmed to execute tasks with precision, often emulating human actions or animal behavior. Integrating principles from mechanical engineering, electronics, computer science, and artificial intelligence, robotics seeks to develop intelligent agents capable of interacting with the physical world, processing sensory data, and making informed decisions. These machines can perform a myriad of functions, from simple actions to intricate manoeuvres in various environments. Robotics finds applications across industries such as manufacturing, healthcare, exploration, and more, enhancing efficiency, safety, and innovation. As technology advances, robotics continues to redefine our interaction with automation, presenting boundless opportunities for advancement and progress.

9.2 MAIN OBJECTIVES

1. Automation: Robotics aims to automate tasks that are repetitive, time-consuming, and labor-intensive, reducing the need for human intervention and increasing efficiency.
2. Precision: Robots are designed to perform tasks with high precision, accuracy, and consistency, minimizing errors and variations in outcomes.

3 Efficiency: The primary objective of robotics is to enhance operational efficiency by streamlining processes, optimizing resource utilization, and reducing waste.
4 Safety: Robots can be employed in hazardous environments where human presence is risky, ensuring the safety of workers while accomplishing tasks efficiently.
5 Productivity: Robotics seeks to boost overall productivity by enabling continuous operations, minimizing downtime, and increasing output rates.
6 Complex tasks: Robots are developed to execute complex and intricate tasks that may be beyond human capabilities, such as heavy lifting, precise assembly, or working in extreme conditions.
7 Cost reduction: By automating various tasks, robotics helps lower labor costs, minimize operational expenses, and improve resource allocation.
8 Exploration and discovery: In fields like space exploration and deep-sea research, robots are deployed to explore environments that are inaccessible to humans, thus facilitating new discoveries.
9 Assistance and rehabilitation: Robotics aims to assist people with disabilities and enhance rehabilitation processes through devices that mimic human movements and aid in physical therapy.
10 Education and research: Robots are utilized in educational settings to promote STEM learning and research, enabling students and researchers to explore various scientific and engineering principles.
11 Innovation: Robotics drives technological innovation by pushing the boundaries of what machines can achieve, encouraging the development of new solutions and capabilities.
12 Human-machine interaction: Robotics seeks to create machines that can interact and collaborate with humans effectively, enhancing cooperation and augmenting human abilities.
13 Exploration and entertainment: Robots are used in fields like underwater exploration, aerial photography, and entertainment to capture unique perspectives and provide engaging experiences.
14 Environmental conservation: Robotics contributes to sustainability efforts by enabling precise and targeted resource management, reducing waste, and optimizing energy usage.
15 Future adaptability: The objective of robotics is to create adaptable and versatile machines capable of evolving to meet changing technological, industrial, and societal needs.

These objectives collectively drive the evolution and advancement of robotics, creating a transformative impact across various sectors and industries [6].

9.3 AGRICULTURE

9.3.1 Labour Crisis

The modern agricultural landscape faces a critical labor crisis that poses significant challenges to the industry's efficiency and sustainability. This crisis stems from a confluence of factors, including a declining workforce interested in pursuing agricultural careers, seasonal demands for labor-intensive tasks, the physical toll of farming tasks leading to injuries and health issues, an aging workforce, global migration trends, and unattractive working conditions. As the shortage of labor becomes more pronounced, the need for innovative solutions becomes increasingly evident.

Robots have emerged as a transformative solution to mitigate the labor crisis in agriculture. By harnessing the power of automation and technology, robots offer a range of advantages that address the challenges posed by workforce shortages. Their efficiency, precision, and ability to work consistently around the clock make them ideal candidates for tasks ranging from planting and fertilization to harvesting and maintenance. The use of robots not only boosts overall productivity but also ensures tasks are completed on time, regardless of seasonal fluctuations or labor shortages.

The integration of robotics in agriculture offers a potent solution to combat the labor crisis and ensure the industry's continued growth. By leveraging automation, precision, and efficiency, robots not only mitigate the challenges posed by workforce shortages but also redefine the way agricultural tasks are executed. Embracing robotics promises enhanced safety, increased productivity, and improved resource utilization, thereby shaping a more prosperous and sustainable future for the agriculture sector.

9.3.2 Technological Developments

Technological advancements have ushered in a new era of innovation in the field of robotics in agriculture. These developments have revolutionized traditional farming practices, offering more efficient, precise, and sustainable solutions. Several key technological trends are driving the transformation of agriculture through robotics:

1. Sensing and perception technologies: Robotics in agriculture heavily relies on sensors, cameras, LiDAR, and other perception technologies to gather real-time data about the environment. These sensors enable robots to accurately detect crops, weeds, pests, and soil conditions. This data is essential for making informed decisions during tasks like planting, irrigation, and pesticide application.
2. Machine learning and AI: Machine learning algorithms and artificial intelligence play a pivotal role in enhancing the capabilities of

agricultural robots. These technologies enable robots to identify and classify different crops, pests, and diseases, allowing for targeted and precise interventions. AI also enables robots to adapt to changing conditions and optimize their actions based on historical and real-time data.

3 Autonomous navigation: Navigation systems powered by GPS, RTK (Real-Time Kinematic), and advanced mapping algorithms enable robots to move autonomously through fields with high precision. Autonomous navigation ensures that robots follow predefined paths, minimizing crop damage and optimizing efficiency.
4 Aerial and drone technology: Drones equipped with cameras and sensors provide a bird's-eye view of the fields, enabling farmers to monitor crop health, growth patterns, and pest infestations. Drones can also be used for precision spraying and planting, covering large areas quickly and accurately.
5 Swarm robotics: The concept of using multiple robots working together as a swarm is gaining traction in agriculture. Swarm robotics enables efficient coverage of fields, collaborative tasks such as pollination, and sharing of data among robots to make informed decisions collectively.
6 Data analytics and decision support: The data collected by robotic systems is processed using advanced analytics tools to provide actionable insights for farmers. These insights help optimize planting patterns, irrigation schedules, fertilization strategies, and pest control measures.
7 Remote monitoring and control: Many robotic systems can be monitored and controlled remotely through smartphones or computers. This capability allows farmers to supervise operations, make adjustments, and receive real-time updates on the progress of tasks being performed by the robots.

These technological developments in robotics have transformed the way farming is conducted, enabling greater precision, efficiency, and sustainability. As technology continues to evolve, the potential for agricultural robots to address challenges and optimize various aspects of farming becomes increasingly promising.

9.3.3 Advancements in Precision Agriculture: Integration of CAN-Based Control Systems, Sensor Technologies, and Automation

In modern agriculture, there is a paradigm shift towards employing advanced control systems and information technologies to optimize farming operations. This is particularly evident in the integration of CAN-based distributed control systems and information networks in agricultural machinery. Nordic farmers are adopting these systems, which serve as a foundational platform for incorporating enhanced control functions and data-driven decision-making. [7]

While GPS localization has revolutionized precision agriculture, it faces challenges in specific agricultural environments. For instance, factors like varying terrains and dense vegetation can hinder accurate GPS-based localization. In response, technologies like Simultaneous Localization and Mapping (SLAM) algorithms are gaining importance. These algorithms enable agricultural machinery to navigate and operate effectively in real-time by combining sensor data and environmental mapping.

When the agricultural operation begins, a digital map of the field is downloaded to the machinery. This map contains essential information about the layout of the crops, soil conditions, and potential obstacles. Throughout the process, sensors on the machinery continuously gather data, such as plant health indicators and soil moisture levels. These measurements aid in making informed decisions about irrigation, fertilization, and other interventions. The machinery employs advanced algorithms to analyze this data, allowing for targeted actions that maximize crop yield and quality.

Moreover, the integration of autonomous drones and robotic vehicles is gaining traction in agriculture. Drones equipped with various sensors provide a bird's-eye view of the fields, allowing farmers to assess crop health, identify areas of concern, and even apply targeted treatments. Robotic vehicles equipped with precise navigation systems and specialized tools can perform tasks like planting, weeding, and harvesting with remarkable accuracy and efficiency (Figure 9.2).

Figure 9.2 Camera View with Rows Identified.

In conclusion, the agricultural sector is undergoing a digital transformation, driven by the integration of CAN-based control systems, sensor technologies, data analytics, and automation. These technologies empower farmers to make informed decisions, optimize operations, and achieve higher productivity and sustainability. Just as in other industries, the digital revolution is reshaping agriculture, paving the way for more efficient, precise, and data-driven farming practices.

9.3.4 Autonomous Navigation and Mapping

In the realm of mobile robotics, the exploration of Simultaneous Localization and Mapping (SLAM) algorithms has been extensive. In the context of agriculture, the primary objective revolves around identifying and characterizing valuable crops amidst various other plants and non-essential vegetation. This task presents considerable challenges due to fluctuating illumination conditions, the presence of bushes, young trees, and branches from valuable crops. Moreover, agricultural operations must continue during winter when heavy snowfall can obscure the crops and require machines to operate in low-light conditions during early mornings and late evenings.

Autonomous SLAM capabilities stand as a pivotal aspect of outdoor mobile robots, a category which includes those employed in agriculture. Navigation in outdoor settings remains a challenge due to the absence of distinct features, necessitating more sophisticated perception and modeling techniques. 2D laser range finders have gained traction as optimal sensors for localization and map construction due to their cost-effectiveness and precision. Laser scanners offer accurate range and bearing information with sub-degree resolution and accuracies in the range of 1–10 cm over distances of 10–50 meters [8].

Groundbreaking work has been undertaken at institutions such as the University of Sydney by researchers like Eduardo Nebot and Tim Bailey, focusing on robot navigation in complex outdoor environments. The Australian National University has also contributed valuable resources related to forestry measurements.

While the vision of fully autonomous agricultural operations remains a distant goal, incremental progress is achievable. Developing a SLAM-based support system for agricultural machinery operators appears to be a feasible and rewarding endeavor. Initiatives like the ongoing project 'Forestrix' at Helsinki University of Technology (TKK) are investigating technologies, signal processing techniques, and algorithms to develop such support systems for agricultural machinery operators. Additionally, exploring semi-autonomous control of agricultural machinery for less demanding conditions is under scrutiny. Advancements in laser range finders and machine vision systems offer novel opportunities for enhancing agricultural measurements and operations.

9.3.5 Advancing Precision with Automatic Guidance

Since the earliest horse-ploughing contests, farmers have sought the efficiency of straight lines in their fields, a concept reinforced by the notion of 'controlled traffic' in countries like Australia. This approach aims to minimize soil disruption by keeping vehicles within established wheel ruts over successive years. To achieve this, an automatic guidance system with centimeter-level accuracy is highly desirable. While options like buried cables were considered, the potential of machine vision and GPS gained prominence in the early 1990s. Researchers in Southern Queensland successfully achieved automatic steering with centimeter accuracy using a simple frame-grabber and color-based image analysis. This progress was fuelled by technological advancements in computing and affordable video capture systems, ultimately leading to successful market adoption.

9.3.6 Precision Operations in Sowing, Weeding, Spraying, and Harvesting

Different agricultural operations demand varying levels of technological precision. For tasks like spraying and yield monitoring, half-meter accuracy is often adequate. Balancing speed and accuracy is crucial, leading to a trade-off between inertial-assisted 5 Hz GPS and higher frequency alternatives. In operations like 'listing up,' where visual cues are limited, precision GPS and inertial sensors prove advantageous. Machine vision excels over GPS in stages such as seedling emergence and cultivating, as it effectively reduces positional errors and optimizes blade placement. When harvesting crops like cotton and wheat, mechanical 'stem feelers' and vision-based boundary detection contribute to precise harvesting techniques [9].

9.3.7 Enhancing Irrigation and Decision-making with Automation

The integration of automation extends beyond traditional crop-related activities to include irrigation systems. Center-pivot irrigation devices, characterized by self-propelled mechanisms, can cover vast fields—up to 600 acres per pass. These systems are further empowered by sensory inputs like GPS, moisture monitors, and imaging devices, enhancing their decision-making capabilities. This allows targeted water and fertilizer application based on real-time field conditions, significantly optimizing resource utilization.

9.3.8 Striking a Balance between Innovation and Practicality

As the realm of automation and precision technologies continues to reshape modern agriculture, broad acre applications stand as a testament to the industry's transformative potential. Achieving equilibrium between cutting-edge innovations and practical implementation remains essential, alongside addressing challenges like regulations and the ongoing pursuit of fully autonomous agricultural machinery. Collaborative efforts between robotics and farming promise a future that embraces technology while honoring the wisdom of centuries-old farming practices.

9.3.9 Intelligent Harvesting: Challenges and Innovations

As we shift away from large-scale crop cultivation, the intricacies of harvesting demand refined selection and enhanced sensing techniques. While manual labor or traditional 'tree shakers' are employed for certain fruits like citrus, the practice of hand picking remains predominant. However, the pursuit of intelligent picking systems has posed a substantial challenge for robotics researchers

9.3.10 Location-Based Picking and Fruit Ripeness Detection

Harvesting operations often involve tasks of pinpointing the optimal location for picking and evaluating fruit readiness. This dynamic task can encompass deriving target positions for picking actuators. Additionally, determining which fruits are ripe for harvest and which require further ripening presents an added complexity. At present, 'automated' solutions involve the integration of conveyor belts alongside tractors, to manually select and harvest crops like broccoli, rockmelons, and cauliflowers.

9.3.11 Automated Sorting

Beyond harvesting, the focus shifts to sorting and grading produce. Some tomato growers opt for mass-harvesting green tomatoes and ripening them in controlled atmospheres. This transition transforms the selection task from field operations to color-sorting lines. Implementing sorting lines involves various actuation mechanisms, from simple tipping systems to air-jets that deflect falling produce into appropriate bins. The challenge of singulation, separating individual items for inspection, requires innovative solutions. Whether it's capturing variations in texture, determining ripeness, or even assessing size, machine vision technologies play a pivotal role in automating sorting and grading operations [10].

9.3.12 Pioneering Precision

In the realm of crop harvesting and sorting, innovation continues to bridge traditional practices with cutting-edge technology. The pursuit of intelligent harvesting systems and automated sorting techniques is reshaping the landscape of agriculture. As robotics researchers collaborate with industry experts, we are witnessing a shift towards seamless automation that not only enhances efficiency but also ensures the quality of harvested produce.

9.4 COOPERATIVE AGRICULTURE SYSTEMS

9.4.1 Human Robot Cooperation

Human-robot cooperation is crucial in agriculture, where some tasks are tough for robots to do alone. There are different ways to make this cooperation work, like remote control, supervisory control, or cooperation. Let's look at examples in agriculture.

Cooperative control needs a connection between the human and the robot, so they can share info and control. For instance, a vineyard sprayer robot can do things by itself, but a human can intervene through a control panel. Another example is using signals from a human's muscle activity to steer a tractor.

A framework for human-robot teamwork was made for citrus harvesting. It showed how humans can help when robots have problems. Robots can also recognize human actions, like lifting a crate, using wearable sensors. They can even spot humans in fields from a distance.

A model was created for humans, robots, sensors, and software to work together. This system had layers for communication, interaction, organization, and collective intelligence. Another study improved pesticide delivery using a human-robot interface. The robot followed a better path after a human showed it how.

In short, human-robot cooperation in agriculture is about improving robot senses and assisting humans with tough tasks. How much robots should do and how much humans should do depends on the task. Creating good interfaces and ensuring safety are important in this teamwork [11].

9.4.2 UAV Robot Teams

Unmanned Aerial Vehicles (UAVs) are used in agriculture for tasks like remote sensing, mapping, monitoring, and pest control. However, using a single UAV has limitations in terms of time efficiency and battery life. To overcome these limitations, researchers have explored the use of multiple cooperating UAVs for tasks in agriculture [12].

One challenge with a single UAV is that it takes a long time to cover a large area, and frequent recharges are needed due to battery constraints. Using a team of UAVs can divide the workload and reduce task duration, which also helps conserve energy. This approach involves coordination of

UAV movements and is particularly important in agriculture due to the nature of the tasks.

Efforts are made to optimize UAV cooperation. A system was designed for precision agriculture

Path planning and coordination algorithms are crucial for effective multi-UAV cooperation. Techniques like area decomposition and spatial interference minimization are used to allocate tasks among UAVs efficiently. Additionally, experiments and simulations help evaluate the performance of these cooperative systems.

In summary, multiple UAVs working together can enhance efficiency and coverage in agriculture. The research primarily revolves around developing algorithms for path planning, coordination, and collision avoidance. As technology advances, future work will likely focus on improving UAV perception capabilities, battery life, and manipulation skills

9.4.3 UGV Robot Teams

Efficient use of multiple robots in farming requires careful planning and optimization of tasks and paths. This is crucial to reduce operational costs. For instance, employing controlled traffic strategies for pairs of agricultural robots compared to uncontrolled traffic was shown to enhance efficiency in terms of distance travelled (Figure 9.3).

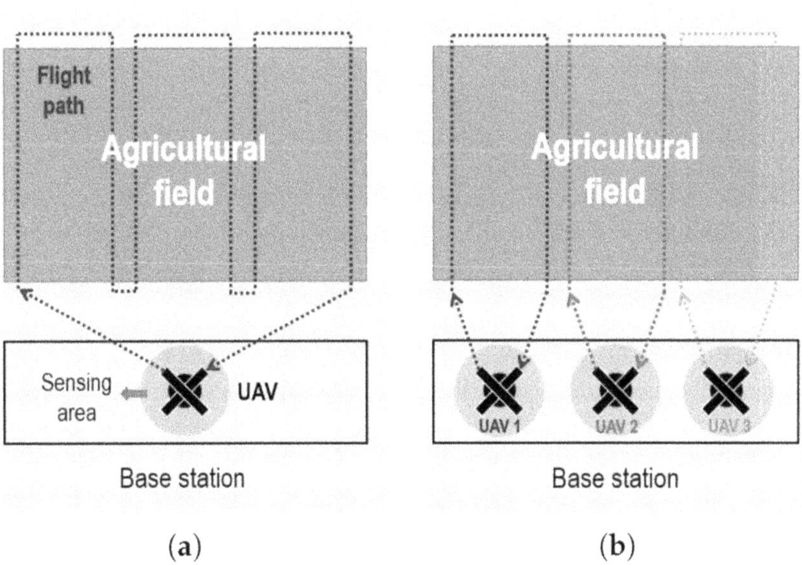

Figure 9.3 UAV Cooperative System.

In contrast to UAVs, UGVs are more suitable for tasks like sowing, spraying, and harvesting. Research in this area focuses on improving task efficiency through multi-robot teams. Effective management of robot allocation and coordination, often using leader-follower approaches, is crucial. There's room for further research to enhance collaborative tasks among ground robots, like load transfer and concurrent harvesting [13].

9.4.4 Cooperative Manipulators

Cooperative manipulation in agricultural robotics involves the coordinated operation of robotic manipulators mounted on the same platform. This approach aims to enhance efficiency and reduce task duration by having multiple arms collaborate towards a common goal, such as harvesting a fruit. The principles of cooperative manipulation can also extend to manipulators mounted on separate platforms.

Several examples of cooperative manipulation systems have been developed for various agricultural tasks:

1. A multi-arm kiwi harvester with four 3 Degrees-of-Freedom (DoF) arms, each equipped with custom grippers, demonstrated independent harvesting by arms. An overall task scheduler assigned fruit clusters to each arm based on detected fruits.
2. A dual-arm strawberry harvesting robot used collision avoidance and a harvesting order planner to control two 5 DoF manipulators. The dual-arm approach demonstrated reduced harvesting times compared to a single manipulator.
3. A system using two manipulators for apple harvesting was designed, where one arm picked apples and the other arm caught them. Different harvesting strategies were evaluated to minimize damage.
4. A tomato harvesting robot employed two mirrored 3-DoF arms with different end-effectors, enabling efficient cutting and grasping of tomatoes.

Overall, most cooperative manipulation systems focus on task assignment and collision avoidance for robotic arms on a single platform. A few studies explore direct cooperation between arms, such as simultaneous fruit picking, which can be advantageous for occluded or complex tasks. Although there's been progress in multi-arm systems, applications involving arms mounted on separate mobile platforms in agricultural robotics are yet to be explored. Such systems could offer greater flexibility, improving efficiency and reducing operational costs.

9.5 CONCLUSIONS AND FUTURE DIRECTIONS

As robotics continues to make its presence felt across various domains, its transformative impact on agriculture is becoming increasingly evident. The applications of robotics in farming are diverse and wide-ranging, with ample room for further innovation and development. The agricultural sector, essential to our society, is facing challenges such as a shortage of skilled labor and the need for more efficient seasonal harvesting operations. Intelligent robotics presents a promising solution to address these challenges and revolutionize the industry.

The integration of robotics into agriculture holds significant promise. However, several critical factors and challenges must be considered as the field progresses:

1. Expertise preservation: The agricultural industry is experiencing a gradual loss of expertise due to demographic shifts and changing trends. Robotic systems must be designed not only to perform tasks but also to facilitate knowledge transfer and assist in training the next generation of farmers.
2. Robust sensors and actuators: Agricultural environments are characterized by harsh conditions and environmental extremes. Future advancements will require the development of rugged sensors and actuators that can endure these challenging working conditions, ensuring reliable and accurate performance of robotic systems.
3. Cost efficiency: Many agricultural operations operate on small profit margins, which presents a unit cost barrier for implementing robotics. To address this challenge, either a generalized methodology for designing and deploying robotic systems across various agricultural tasks needs to be established, or alternative measures to reduce unit costs must be explored.
4. Collaboration and interoperability: As robotic technologies become more prevalent in agriculture, ensuring compatibility and interoperability between different robotic platforms, systems, and equipment will be crucial. Collaboration between research institutions, industry stakeholders, and regulatory bodies is necessary to establish standardized interfaces and protocols.

Looking ahead, the future of agricultural robotics is full of possibilities. Research and development efforts should continue to focus on designing intelligent robotic systems that not only enhance efficiency but also promote sustainability, reduce environmental impact, and contribute to global food security. Collaborative efforts between academia, industry, and policymakers will be instrumental in shaping the trajectory of agricultural robotics, making it a vital part of modernizing and advancing this crucial industry. With continued innovation and interdisciplinary collaboration, robotics will undoubtedly play a pivotal role in shaping the future of agriculture.

REFERENCES

1. Johnson, A. W. & Earle, T. 2000. "The evolution of human societies," From Foraging Group to Agrarian State, Second Edition, **53**.
2. *Sustainable Agricultural Mechanization* (PDF). Rome: Food and Agriculture Organization of the United Nations (FAO), 2016. https://openknowledge.fao.org/server/api/core/bitstreams/e9bffdc7-1dbc-4354-ab2b-eb316c71ab81/content
3. Eliazer Nelson, A. R. L.; Ravichandran, K.; Antony, U. (1 October 2019). "The impact of the Green Revolution on indigenous crops of India". *Journal of Ethnic Foods*. **6** (1): 8. doi:10.1186/s42779-019-0011-9. ISSN 2352-6181. S2CID 204873993.
4. Wolfert, S.; Ge, L.; Verdouw, C.; Bogaardt, M.-J. (1 May 2017). "Big data in smart farming – A review". *Agricultural Systems*. **153**: 69–80.
5. Foglia, M. M.; Reina, G. (2006). "Agricultural robot for radicchio harvesting." *Journal of Field Robotics*. **23** (6–7): 363–377. doi:10.1002/rob.20131.
6. Roozing, W.; Li, Z.; Tsagarakis, N.; Caldwell, D. (2016). "Design optimisation and control of compliant actuation arrangements in articulated robots for improved energy efficiency". *IEEE Robotics and Automation Letters*. **1** (2): 1110–1117.
7. Cena, G.; Valenzano, A. (1999). "Overclocking of controller area networks". *Electronics Letters*. **35** (22): 1923. Bibcode: 1999ElL....35.1923C. doi:10.1049/el:19991289
8. Julier, S.; Uhlmann, J. (2001). "Building a million-beacon map." *Proceedings of ISAM Conference on Intelligent Systems for Manufacturing*. SPIE Digital Library Boston, MA, United States doi:10.1117/12.444158
9. *The State of Food and Agriculture 2022 – Leveraging agricultural automation for transforming agrifood systems*. Rome: Food and Agriculture Organization of the United Nations (FAO), 2022. doi:10.4060/cb9479en. ISBN 978-92-5-136043-9.
10. Pla, F.; Sanchiz, J. M.; Sanchez, J. (2001). "An integral automation of industrial fruit and vegetable sorting by machine vision." – *Proceedings of 8th IEEE International Conference on Emerging* , IEEE, Antibes-Juan les Pins, France
11. Lytridis, C., Kaburlasos, V. G., Pachidis, T., Manios, M., Vrochidou, E., Kalampokas, T., & Chatzistamatis, S. (2021). "An overview of cooperative robotics in agriculture. *Agronomy* **11** (9): 1818. doi:10.3390/agronomy11091818

12. Mammarella, M., Comba, L, Biglia, A., Dabbene, F., Gay, P. Cooperation of unmanned systems for agricultural applications: A theoretical framework., Biosystems Engineering, Volume 223, 2022, Pages 61-80 doi:10.1016/j.biosystemseng.2021.11.008
13. Grocholsky, B.; Keller, J.; Kumar, V.; Pappas, G. (2006) "Cooperative air and ground surveillance." *IEEE Robotics and Automatic Magazine.* **13** (3): 16–25.

Chapter 10

A Smart Irrigation System for Efficient Water Requirements Using Weather Prediction

Riya Parikh

10.1 INTRODUCTION

10.1.1 Importance of Technology in Agriculture

Technology has played a key role in transforming the agricultural sector, leading to substantial advancements and improvements. The use of technology in agriculture can be proven to be very useful for the following reasons:

- Increased productivity from innovations such as precision agriculture, automated machinery, and sensor-based systems.
- By monitoring soil health, water usage, and pest infestations through data-driven techniques, farmers can adopt sustainable techniques for maximizing production.
- Smart farming technologies, like IoT devices and data analytics, help in the precise management of resources such as water, fertilizers, and pesticides.
- Advanced agricultural technologies assist in enhancing the quality of crops through better pest control, disease monitoring, and harvesting practices.
- Weather forecasting tools assist farmers in anticipating extreme weather events, enabling them to take necessary measures to protect their crops and livestock. This decreases the risk of crop failure and financial losses.

10.1.2 Scarcity of Water and Irrigation Systems Available

Water management is also a crucial problem in the agriculture sector as it uses 70% of total water usage globally. As the world's population surges, the demand for water rises. By 2025, 46 to 52 countries will face water

 DOI: 10.1201/9781003676102-10

scarcity. Various smart irrigation techniques have been adopted in India for effective water management:

- Drip irrigation
- Sprinkler irrigation
- Soil moisture sensors
- Weather-based irrigation controllers
- Solar-powered irrigation
- Water recycling and reuse system.

These methods have some benefits one over others and are chosen as per the requirements. The water level needing to be maintained for the crops differs depending on the species, the stage of growth, and the environmental conditions. If crops do not receive the required amount of water, they can droop and become stressed, inhibiting their growth and even death. Conversely, if plants receive excess water, it can lead to root rot and other fungal diseases, harming or killing the plant. The water level needed for crop development depends on factors such as soil type, climate, and the specific needs of the plant species [1].

Many researchers have proposed different models to address the above mentioned challenges. In this chapter we have proposed a model, in which crop water needs have been predicted based on species, evapotranspiration rate, soil type, and growth stage. The details of this are provided further in this chapter. Alongside, rainfall was also predicted using machine learning. With the help of machine learning, rainfall prediction has become more accurate and reliable, providing valuable insights for decision-making. Machine learning algorithms can analyze huge amounts of historical data to identify patterns and correlations between different weather variables, such as temperature, humidity, wind speed, cloud cover, and solar radiation. It can also take climate change into account. This allows the algorithm to make predictions based on the current weather conditions and historical data. These predictions were used with the data from soil moisture sensors to predict the water needed to be given from external sources. The water level was efficiently maintained with the use of PID controllers.

10.2 LITERATURE REVIEW

The current era necessitates technological progress to enhance convenience in daily life. The acceleration of technological domains like robotics, artificial intelligence, and the Internet of Things (IoT) is also proving advantageous in the agricultural sector. This section elucidates various endeavors undertaken by researchers in the realm of intelligent irrigation systems.

10.2.1 Yearwise Research

10.2.1.1 2023

One such system, developed by Kumar et al. [2], utilizes a cutting-edge GSM system and a sensitive soil moisture sensor for automated irrigation control. These cutting-edge systems are engineered to offer smooth and automated oversight of irrigation procedures through the utilization of a state-of-the-art valve mechanism.

10.2.1.2 2022

Pramanik et al. [3] improved irrigating effectiveness by up to 86.6% significantly using the soil moisture-based capacitance method. Obaideen et al. [4] demonstrated that IoT integration in agriculture aligns with the sustainable development goals. Ndunagu et al. [5] utilized WSN and the ThingSpeak platform for smart irrigation and water forecasting. Sensor information was transmitted every 15 minutes following detection. The open cloud platform referred to as ThingSpeak was employed to predict crop water requirements. Within the realm of the Internet of Things (IoT), LPWAN finds widespread usage, especially within agriculture, due to its energy-efficient attributes that effectively supported sensor nodes situated in open agricultural spaces. Mekki et al. [6] explored the potential of LPWAN solutions like Sigfox, NB-IoT, and LoRaWAN for IoT advancements in agriculture.

10.2.1.3 2021

Kassanuk et al. [7] implemented an IoT-based SIS with cloud-based soil moisture storage to control sprinkler irrigation. Sharma et al. [8] determined the water requirements of rice at different growth stages using water flow sensors and Node MCU. Patel et al. [9] developed a water-saving model for farms using shared water resources.

10.2.1.4 2020

Velmurugan [10] predicted rainfall using weather data and measured soil moisture to trigger irrigation. Vij et al. [1] used WSN to collect sensor data for crop analysis in specific regions of India, achieving accurate water prediction with K-NN. Reddy et al. [11] employed decision tree algorithms for water supply forecasting based on temperature, humidity, and soil moisture data.

10.2.1.5 2019

Laksiri et al. [12] used weather data prediction and soil moisture sensors for irrigation scheduling.

10.2.1.6 2018

Goap et al. [13] introduced a system that gathers weather information from a weather app to predict upcoming weather conditions for the next 34 days. This system integrates sensors for soil moisture, temperature, and humidity to enhance the precision of forecasts. The measurement of soil moisture content is achieved through the utilization of time-domain reflectometry (TDR).

10.2.1.7 2016

Arvindan et al. [14] designed a system using Arduino-UNO, HC-05 Bluetooth, and a smartphone for soil moisture-based irrigation scheduling. Rajalakshmi and Mahalakshmi [15] developed a smart irrigation architecture. Davis and Dukes [16–17] put forward an intelligent irrigation system dependent on weather conditions, capable of functioning effectively with moderately textured soil across various crop varieties. The entire control of this system is managed through a mobile application. This system acquires real-time weather information from a weather app, calculates EvapoTranspiration (ET) using historical data, and subsequently adjusts the irrigation operations in response. Ranade and Takale [18] used FPGA units and Zigbee for communication in their system.

10.2.1.8 2015

Harun et al. [19] designed an XBee-based automated irrigation system for greenhouses with temperature, humidity, and soil moisture sensors. Similarly, Chikankar et al. [20] devised an automated irrigation setup that harnessed metrics like temperature, soil moisture, and air humidity. These variables held vital significance in the context of Precision Agriculture (PA) control. The system operation was governed by an Arduino Uno IoT controller, affording farmers the capability to oversee sensor data remotely from their residences. The system was engineered to accommodate diverse crop categories like wheat, rice, jowar, and bajara within a master node, enabling sequential irrigation for individual crops per user. The system architecture was configured to cater to any crop type grown on medium-grade soil and was overseen through a mobile application.

10.2.1.9 2014

Nandurkar et al. [21] introduced a system that incorporated signal conditioning circuits for sensor integration, transceiver circuits, as well as temperature and soil moisture sensors. Notably, the system was tailored to cater to a singular crop during its operation. Notably, the system solely relied on real-time determinations and did not involve data storage or the transmission of notifications.

10.2.1.10 2011

Dubey et al. [22] devised a system that employed a wireless sensor network in conjunction with an embedded technique involving DTMF (Dual Tone Multiple Frequency) signalling. This combination was utilized to manage water flow for irrigation in segmented, sprinkler, or drip sections. The smartphone's camera was harnessed to capture images of crops and soil, aiding in the identification of wet or dry soil conditions. Depending on the targeted moisture level, the system activated or deactivated the valve. The system's validation was performed on a pumpkin crop during testing.

10.3 PROPOSED MODEL

The proposed model is based on selection of the algorithm, prediction of rainfall, determination of water requirement and implementation of an automated system of watering.

10.3.1 Different Machine Learning Algorithms

Three different algorithms were compared, and the one with the minimum error was selected from these algorithms. The algorithms compared were linear regressor, multiple regressor, and XGBoost regressor.

Linear regressor: Linear regression fits a linear model with weights to minimize the cost function in the dataset and the targets predicted by the linear approximation [23].

Multiple regressor: This model optimizes the cost function using stochastic gradient descent [24].

XGBoost regressor: XGBoost uses decision trees as the base learners and updates their weights in each iteration based on the gradient descent algorithm, and this helps in handling large-scale data, parallelizing computations, and preventing overfitting.

The RMSE obtained from the given algorithms is as shown in Table 10.1.

10.3.2 Prediction using XGBoost Regressor

The depicted Figure 10.1 illustrates the operational mechanism of the precipitation prediction model. Input parameters encompassed temperature, dew point, humidity, wind speed, cloud cover, and solar radiation. To forecast the precipitation level, an XGBoost regressor was employed. Systematic trials were conducted to optimize its parameters. The final selection entailed 200 n_estimators and a learning rate of 0.1, resulting in a minimal Root

Table 10.1 RMSE of Different Algorithms

Algorithm	*RMSE*
Linear regressor	9.03
Multiple regressor	10.43
XGBoost regressor	8.53

From the experiment, the XGBoost Regressor was used.

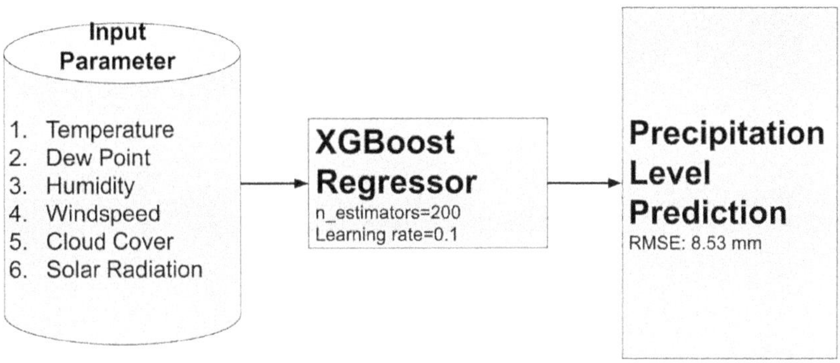

Figure 10.1 Working of Precipitation Model Using XGBoost Regressor.

Mean Square Error (RMSE) on the cross-validation dataset, with a recorded minimum RMSE of 8.53 mm/day.

10.3.3 Overall Working of the Rainfall Prediction Model

Weather data was taken from the source. Data pre-processing was applied to the data. An algorithm was selected based on minimum RMSE. Rainfall was predicted based on the XGBoost regressor selected and was classified into five categories based on different rainfall levels (Figure 10.2).

10.3.4 Calculation of Water Requirement

Water requirements for rice and maize crops were calculated, and the experiment took place in the year 2022 in Kharagpur, West Bengal, for the growing season of June to September using the following formula:

$$ET_{Crop} = ET_0 \times K_C$$

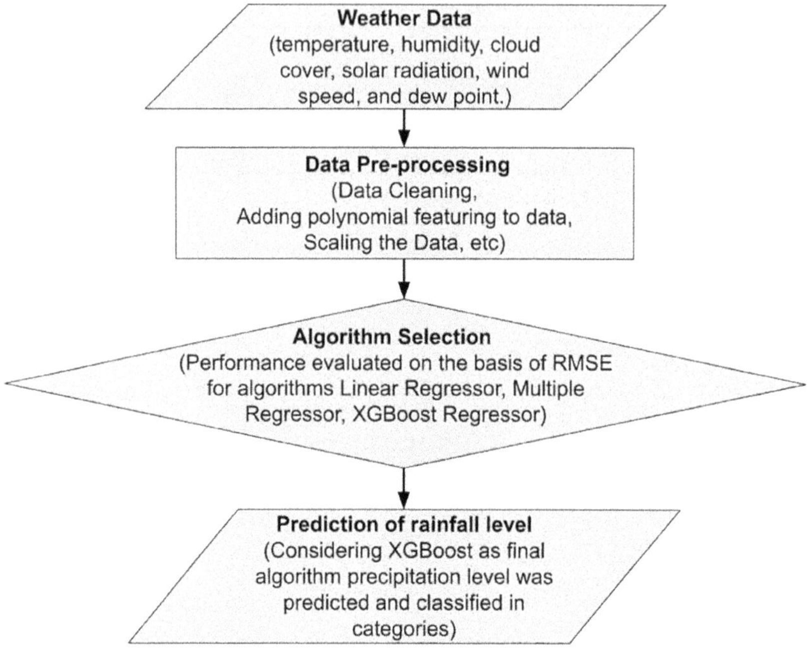

Figure 10.2 Working of the Precipitation Model.

Here, ET_0 is the reference crop evapotranspiration rate, and Kc is the crop factor of the crop.

10.3.4.1 Explanation of Terms used for Calculation

Major Climate factors such as sunshine, temperature, humidity, and wind speed affect the crop water needs. These climatic factors can be considered using the reference crop evapotranspiration (ET_0). The grass is taken as a reference crop [25].

Now water needs vary according to the crop type. This was considered by K_c (crop factor). Every crop has its own crop factor determined experimentally.

The product of these two terms gives ET_{Crop} evapotranspiration of crop in mm/day, which is crop water need.

The crop factor taken for the rice crop and the maize crop is as shown in Tables 10.2 and 10.3.

Table 10.2 Crop Factor of Paddy Crop According to Different Conditions

Climate	*Little wind*		*Strong wind*	
Growth stage (days)	*Dry*	*Humid*	*Dry*	*Humid*
0–60 days after transplant or direct sowing	1.1	1.1	1.1	1.1
Mid-season	1.2	1.05	1.35	1.3
Last 30 days before harvest	1.0	1.0	1.0	1.0

Source: Ref [25].

Table 10.3 Crop Factor of Maize Crop According to Different Conditions

Growth stage (days)	*Crop Factor*
Initial stage (first 20 days)	0.4
Crop development stage (next 25–30 days)	0.8
Mid-season stage (next 25–50 days)	1.15
Late-season stage (next 10 days)	1.00

Source: Ref [25].

This data of water required was used with rainfall prediction to calculate the amount of water needed to be supplied by external sources.

10.3.5 Automated System Developed for Watering

The precipitation level was divided into five categories according to precipitation level:

0–15 mm/day: Trickling
15–30 mm/day: Light rain
30–45 mm/day: Moderate rain
45–60 mm/day: Heavy rain
> 60 mm/day: Very heavy rain

The accuracy of the model, 91.89% and 92% for the 70:30 ratio and 80:20 ratio, respectively, was based on this division.

Watering was undertaken using the Table 10.4 according to the rainfall prediction

This works as follows: for example, if there is heavy rain predicted after 3–4 days, then in this 3–4 days only 50% of the needed water level will be maintained for the crops. And on the day of rainfall predicted no water will be given so that there will be no harm caused to the crops due to heavy rain.

Table 10.4 Percentage Level Maintained According to Rainfall Intensity

Rainfall intensity	*Percentage of needed level maintained*	*Number of day(s) for which given watering level was taken into accounted*
Trickling	95%	1
Light rain	85%	1–2
Moderate rain	75%	2–3
Heavy rain	50%	3–4
Very heavy	40%	4–5

10.3.5.1 How the Water Level was Maintained?

Water level calculated was maintained using the PID Controller method. In specific scenarios, such as chemical and industrial operations, maintaining a consistent liquid level within a tank or similar enclosure holds significance. A control system based on a PID (Proportional-Integral-Derivative) mechanism is used in such cases. This works on a feedback mechanism. This system regulates the water level by modulating the inflow rate of water into the container through the manipulation of the speed of a DC motor-driven water pump, which facilitates the filling process [26].

Figure 10.3 demonstrates the working of the PID controller by feedback mechanism. Here *e*(t) is the error term which is calculated using the: 'level desired-present level formula. The present level of water was determined using a soil moisture sensor.

This works as follows: the error is given as input to the PID Controller. Proportion, integral and derivative terms are calculated and summed up. The level is changed using this PID controller to minimize the error. Again the new error is fed, and the cycle continues until the water level is maintained [27].

The parameters of the PID Controller were then tuned using a program that tune it using Ziegler–Nichols Tuning Method.

The Ziegler Nichols procedure works as follows:

1 *K*p, *K*i, and *K*d were set to 0.
2 A high value for *K*p is assumed until the system is in a steady oscillation around the setpoint. This is known as the critical gain of the system and is denoted as *K*u.
3 The oscillation period is known as *T*u.

Then values shown in Table 10.5 were used for *K*p, *K*d, and *K*i:

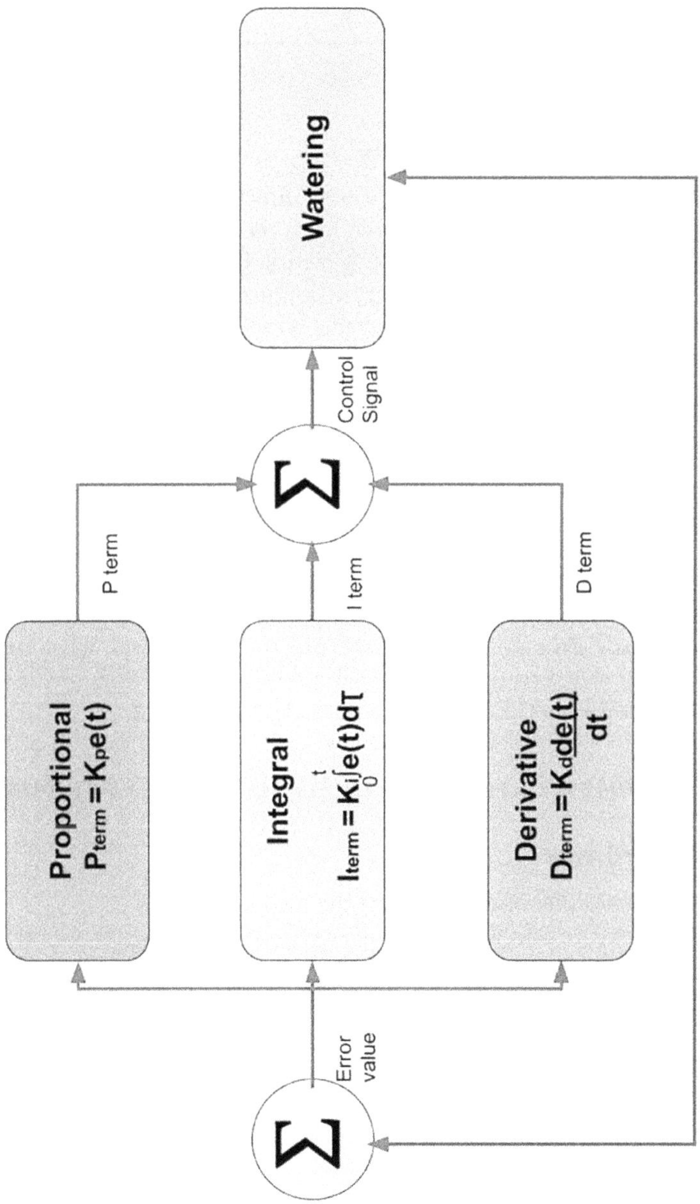

Figure 10.3 Working of PID Controller.

Table 10.5 Values of *Kp*, *Ki* and *Kd*

Kp	*Ki*	*Kd*
0.6Ku	1.2Ku/Tu	0.075*Ku*Tu

Source: Ref [28].

10.3.6 Overall Working Model

Figure 10.4 illustrates the comprehensive functionality of the model. The data utilized in this study was sourced from the open-meteo API [23].

To determine the crop's EvapoTranspiration Rate (ET), we employed ET_0 data from the aforementioned source, in conjunction with the crop's specific *Kc* value for rice and maize. The *Kc* value was selected based on the relative humidity and windspeed information extracted from the data, along with the crop's current growth stage.

The prediction of precipitation levels was accomplished through an Artificial Neural Network (ANN) model. This model considered various factors including temperature, dew point, humidity, windspeed, cloud cover, and solar radiation, all sourced from the API data.

The predicted precipitation level was then used to determine rainfall intensity. This, along with the calculated ET, was fed into the PID controller. PID Controller uses this data to calculate the required water level and soil moisture sensor data to determine the present water level. This was then used to decide when to turn on and off a motor and by how much flow of water. The motor was controlled using an Arduino board.

10.4 EQUIPMENT AND DATA USED FOR THE EXPERIMENT

10.4.1 Hardware

1. Relay module: For controlling the motor
2. Battery: For providing the power required to drive the motor
3. Motor: For irrigating using the water pump
4. Arduino board: For programming the entire circuit
5. Soil moisture sensor: For getting data about soil moisture.

10.4.2 Software (Libraries used)

1. Matplotlib: For plotting the graphs
2. Numpy: For arrays
3. Pandas: For organizing data in the program
4. Sklearn: For linear regressor and multiple regressor and also scaling data and adding polynomial features
5. XGBoost: For XGBoost regressor

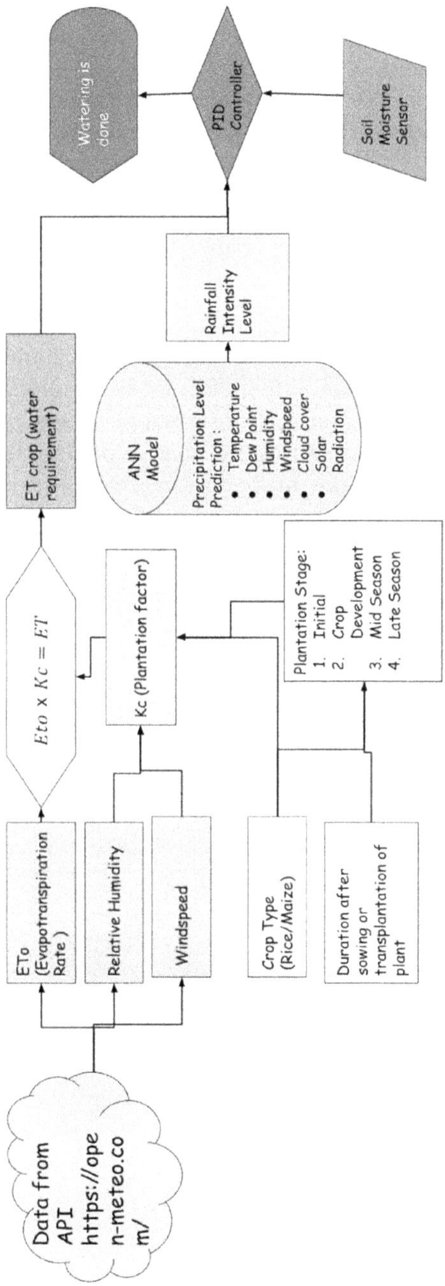

Figure 10.4 Overall Working of Model.

10.4.3 Circuit Diagram of Model

The circuit shown in Figure 10.5 was only for experimental purposes on a small scale. The motor was connected to the battery through a relay module controlled by the Arduino board. Data about soil moisture was taken by a soil moisture sensor, also connected to the Arduino board. Soil moisture needs to be calibrated after connecting it, setting it to 0 for dry soil.

10.4.4 Data Preparation

Prediction of precipitation level had been achieved based on data of 13 years from visual crossing api [29].

Firstly for the prediction, the preparation of data was undertaken. Weather parameters that influence rainfall, like temperature, humidity, wind speed, cloud cover, and solar radiation, had been taken for prediction.

By adding polynomial features to the original dataset, the prediction was made more accurate. For selecting the degree of the polynomial a graph between RMSE, namely, root mean square error in rainfall level against the degree of the polynomial was plotted for the training dataset and cross-validation dataset (FIgure 10.6). The degree with the minimum error of cross-validation dataset was selected as it considers overfitting as well [30].

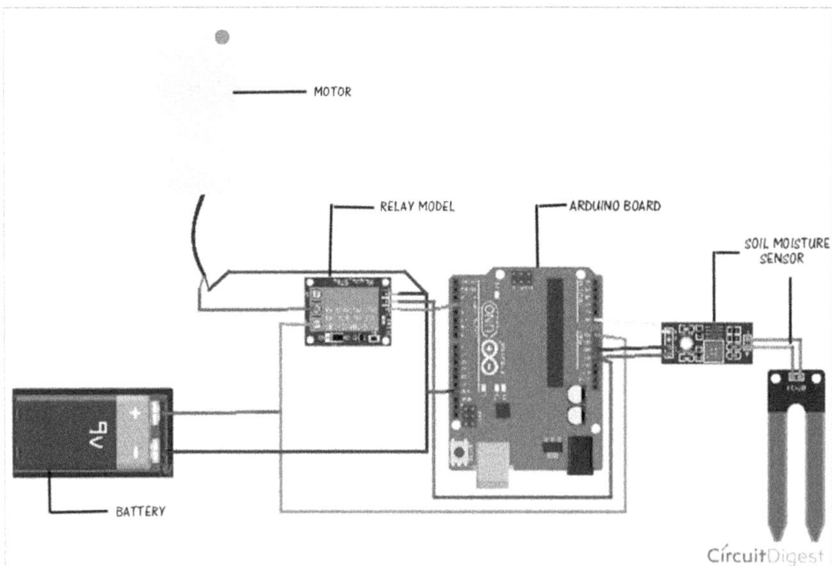

Figure 10.5 Circuit of Model.

Then feature scaling was done to this updated dataset to get processed data for good results. Feature scaling is the process of normalizing the range of features in a dataset, as this dataset contains various features that vary in degrees of magnitude, range, and units [31]. Therefore, in order for machine learning models to interpret these features on the same scale, we need to perform feature scaling.

10.5 RESULTS ANALYSIS

The different algorithms were compared in order to forecast the precipitation level and based on a comparison of RMSE, we finalized the XGBoost regressor algorithm. Systematic trials were conducted to optimize its parameters. The final selection entailed 200 n_estimators and a learning rate of 0.1, resulting in a minimal Root Mean Square Error (RMSE) on the cross-validation dataset, with a recorded minimum RMSE of 8.53 mm/day. The requirement for water is calculated as discussed in Section 3.4 Calculation of Water Requirement

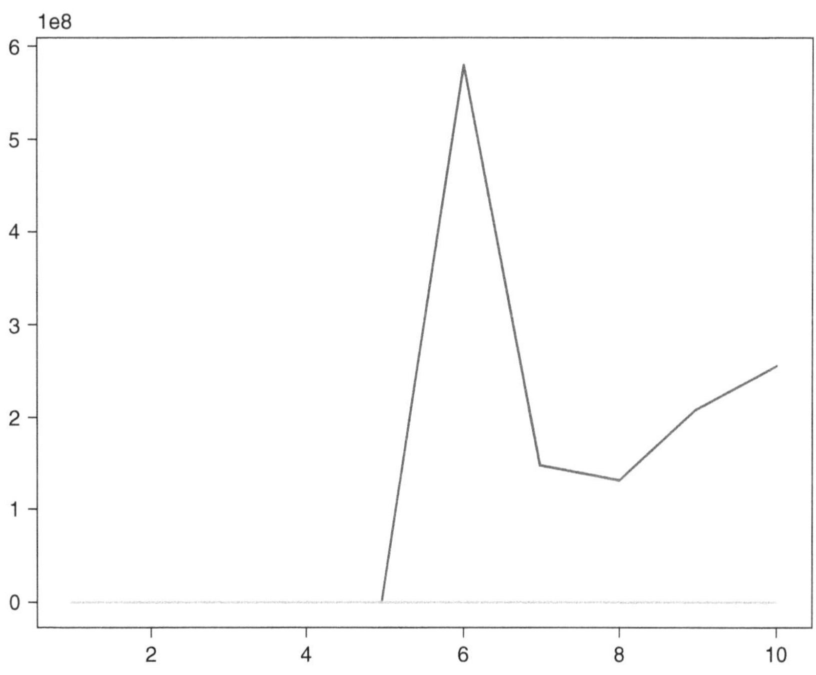

Figure 10.6 Graph of RMSE to the Degree of the Polynomial for Polynomial Features (dark line for Training set and grey line for Cross Validation Set).

$$ET_{Crop} = ET_0 \times K_C$$

This showed a rousing result; on average, only 2.71% of the water needed to be supplied for rice crops and 1.52% for maize crops.

Figure 10.7 shows result for original and predicted data using XGBoost regressor.

This dataset was verified by dividing it into ratios of training and testing, 70:30 and 80:20. The training dataset was further divided into 80:20 ratios for training and evaluation purposes, namely, cross-validation and testing in the XGBoost algorithm for both cases. This gave an accuracy of 91.89% and 92%, respectively.

10.6 CONCLUSION

The novel machine learning framework aims to ascertain the necessary water supply by leveraging weather data. Over a span of 13 years, daily weather information [32] was collected from https://visualcrossing.com/. This model was meticulously developed using Python, and through rigorous experimentation, an impressive 92% accuracy was achieved using the XGBoost algorithm, which outperformed other methods attempted. Around 98% of the water required to be given by irrigation was saved.

This accomplishment opens the door for future advancements in this field, particularly through the incorporation of supplementary data and its application to diverse crops. The model's applicability shines in equatorial regions and Southeast Asia, where frequent precipitation patterns prevail, making it an invaluable asset.

10.7 LIMITATIONS OF THE MODEL PROPOSED

1. The study exclusively focuses on a single season, spanning from June to September.
2. The predictions in this research exclusively utilize weather metrics: temperature, humidity, wind speed, cloud cover, and solar radiation.
3. The scope of predictions is limited to a singular geographic region, namely Kharagpur, West Bengal.
4. The predictions were derived from a dataset spanning 13 years, owing to data availability constraints as outlined in reference [30].

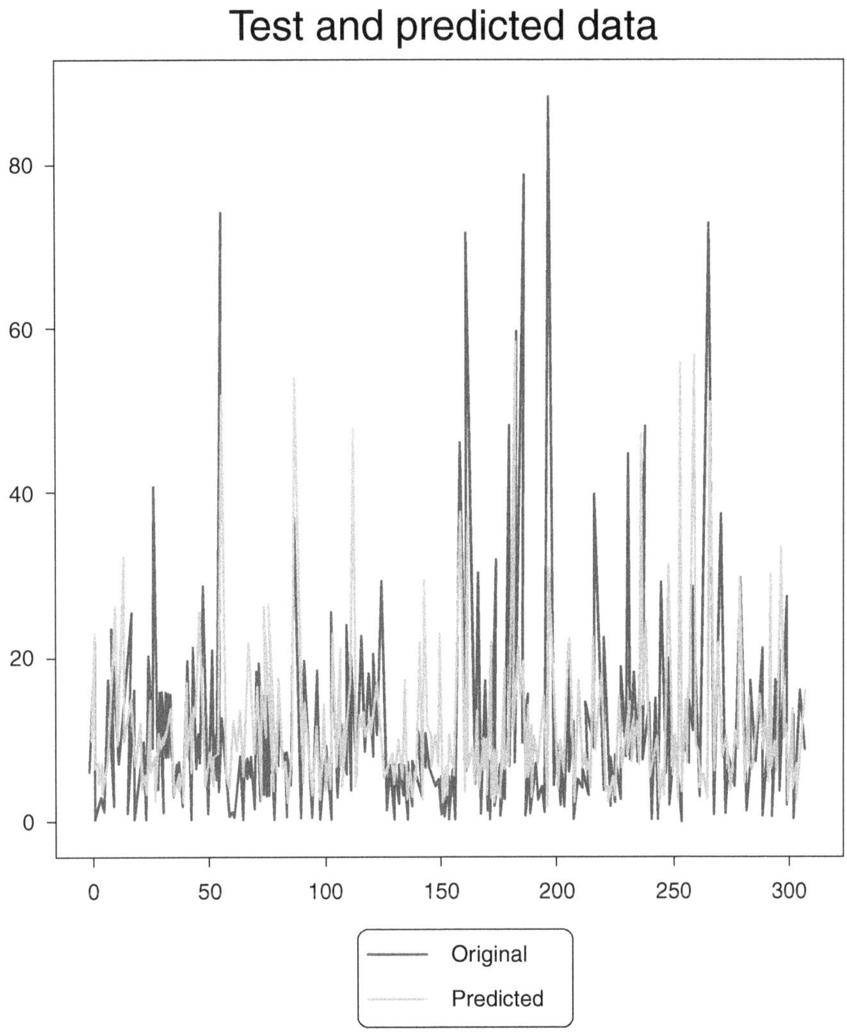

Figure 10.7 Graph of Rainfall Level in (mm/day) to Number of Days.

10.8 FUTURE SCOPE

1 This model can be extended to other crops by taking the data of crop factor of the crop for which prediction is to take place.
2 The prediction can be extended to other seasons also.
3 Other weather parameters on which rainfall may depend can also be used for predicting precipitation levels.

4 Other machine learning models like AdaBoost can also be tested to get better results.
5 This prediction can also be extended to other parts of the world.
6 As for data preprocessing, only polynomial featuring and data scaling has been applied; further researchers can apply other preprocessing techniques.
7 The results of other algorithms than the linear regressor, multiple regressor, and XGBoost can be investigated.
8 A more efficient method to tune the PID controller can be used.
9 More assured ways of taking into account rainfall prediction can be considered.
10 The model has been tested on a small scale by the Arduino board, but it can be extended to a large scale by using other electronic devices like Rasberry Pie.

REFERENCES

1. Vij, A., Vijendra, S., Jain, A., Bajaj, S., Bassi, A., & Sharma, A. (2020). IoT and machine learning approaches for automation of farm irrigation system. *Procedia Computer Science*, 167, 1250–1257.
2. Kumar, N. D., Pramod, S., & Sravani, C. (2023). Intelligent irrigation systems. In *AI for Big Data-Based Engineering Applications from Security Perspectives* (pp. 187–213), eds. Balwinder Raj, Brij B. Gupta, Shingo Yamaguchi, Sandeep Singh Gill. CRC Press.
3. Pramanik, M., Khanna, M., Singh, M., Singh, D. K., Sudhishri, S., Bhatia, A., & Ranjan, R. (2022). Automation of soil moisture sensor-based basin irrigation system. *Smart Agricultural Technology*, 2, 100032.
4. Obaideen, K., Yousef, B. A., AlMallahi, M. N., Tan, Y. C., Mahmoud, M., Jaber, H., & Ramadan, M. (2022). An overview of smart irrigation systems using IoT. *Energy Nexus*, Vol 7, 100124.
5. Ndunagu, J. N., Ukhurebor, K. E., Akaaza, M., & Onyancha, R. B. (2022). Development of a wireless sensor network and IoT-based smart irrigation system. *Applied and Environmental Soil Science*, pp. 1–13, 2022.
6. Mekki, K., Bajic, E., Chaxel, F., & Meyer, F. (2019). A comparative study of LPWAN technologies for large-scale IoT deployment. *ICT Express*, 5(1), 1–7.
7. Kassanuk, T., Mustafa, M., & Panse, P. (2021). An Internet of Things and cloud based smart irrigation system. *Annals of the Romanian Society for Cell Biology*, Vol 25, 20010–20016.
8. Sharma, B. B., & Kumar, N. (2021). IoT-based intelligent irrigation system for paddy crop using an internet-controlled water pump. *International Journal of Agricultural and Environmental Information Systems (IJAEIS)*, 12(1), 21–36.
9. Patel S.V, Parikh, S., & Patel, S. (2021, September). Irrigation to smart irrigation and tube well users. In *2021 International Conference on Computing, Communication and Green Engineering (CCGE)* (pp. 1–5). IEEE.

10. Velmurugan, S. et al., (2020). *An IOT Based Smart Irrigation System using Soil Moisture and Weather Prediction, International Journal of Engineering Research & Technology (ijert) eclectic, Volume 8 – Issue 07, pp. 1–4.*
11. Reddy, K. S. P., Roopa, Y. M., LN, K. R., & Nandan, N. S. (2020, July). IoT based smart agriculture using machine learning. In *2020 Second International Conference on Inventive Research in Computing Applications (ICIRCA)* (pp. 130–134). IEEE.
12. Laksiri, H. G. C. R., Dharmagunawardhana, H. A. C., & Wijayakulasooriya, J. V. (2019, December). Design and optimization of IOT based smart irrigation system in Sri Lanka. In *2019 14th Conference on Industrial and Information Systems (ICIIS)* (pp. 198–202). IEEE.
13. Goap, A., Sharma, D., Shukla, A. K., & Krishna, C. R. (2018). An IoT based smart irrigation management system using Machine learning and open source technologies. *Computers and Electronics in Agriculture*, 155, 41–49.
14. Arvindan, A. N., & Keerthika, D. (2016, March). Experimental investigation of remote control via Android smart phone of arduino-based automated irrigation system using moisture sensor. In *2016 3rd International Conference on Electrical Energy Systems (ICEES)* (pp. 168–175). IEEE.
15. Rajalakshmi, P., & Mahalakshmi, S. D. (2016, January). IOT based crop-field monitoring and irrigation automation. In *2016 10th International Conference on Intelligent Systems and Control (ISCO)* (pp. 1–6). IEEE.
16. Davis, S. L., & Dukes, M. D. (2015). Methodologies for successful implementation of smart irrigation controllers. *Journal of Irrigation and Drainage Engineering*, 141(3), 04014055.
17. Dukes, M. D., Shedd, M. L., & Davis, S. L. (2009). Smart irrigation controllers: Operation of evapotranspiration-based controllers: AE446/AE446, 2/2009. *EDIS*, 2009(2).
18. Ranade, P., & Takale, S. B. (2016). Smart irrigation system using FPGA based wireless sensor network. *International research journal of Engineering and Technology*, 3, 2229–2232.
19. Harun, A. N., Kassim, M. R. M., Mat, I., & Ramli, S. S. (2015, May). Precision irrigation using wireless sensor network. In *2015 International Conference on Smart Sensors and Application (ICSSA)* (pp. 71–75). IEEE.
20. Chikankar, P. B., Mehetre, D., & Das, S. (2015, January). An automatic irrigation system using ZigBee in wireless sensor network. In *2015 International Conference on Pervasive Computing (ICPC)* (pp. 1–5). IEEE.
21. Nandurkar, S. R., Thool, V. R., & Thool, R. C. (2014, February). Design and development of precision agriculture system using wireless sensor network. In *2014 First International Conference on Automation, Control, Energy and Systems (ACES)* (pp. 1–6). IEEE.
22. Dubey, V., Dubey, N., & Chouhan, S. S. (2011, June). Wireless sensor network based remote irrigation control system and automation using DTMF code. In 2011 *International Conference on Communication Systems and Network Technologies* (pp. 34–37). IEEE.
23. MLP Regressor, https://scikit-learn.org/stable/modules/generated/sklearn.neural_network.MLPRegressor.html (29th Feb 2024).

24. https://scikitlearn.org/stable/modules/generated/sklearn.neural_network.MLPRegressor.html (29th Feb 2024).
25. Crop Water Needs, www.fao.org/3/s2022e/s2022e07.htm (29th Feb 2024).
26. Getu, B. N. (2016). Water level controlling system using PID controller. *International Journal of Applied Engineering Research*, 11, 11223–11227.
27. Cataldo, A., De Benedetto, E., Cannazza, G., Piuzzi, E., & Pittella, E. (2017). TDR-based measurements of water content in construction materials for in-the-field use and calibration. *IEEE Transactions on Instrumentation and Measurement*, 67(5), 1230–1237.
28. Tuning Methods of a PID Controller, www.ctrlaltftc.com/the-pid-controller/tuning-methods-of-a-pid-controller (29th Feb 2024).
29. Global Weather Data & API for Every Application, www.visualcrossing.com (29th Feb 2024).
30. Polynomial Features, https://scikit-learn.org/stable/modules/generated/sklearn.preprocessing.PolynomialFeatures.html (29th Feb 2024).
31. Standard Scalar, https://scikit-learn.org/stable/modules/generated/sklearn.preprocessing.StandardScaler.html(29th Feb 2024).
32. Free Weather API, www.open-meteo.com (29th Feb 2024).

Chapter 11

Drones

Their Impact on Agriculture

Vedaant Agarwal

11.1 INTRODUCTION

Agriculture has been the backbone of civilization for thousands of years. It has a key contribution to feeding the world's population, lowering poverty rates, and the global economy. According to the Statista, from 2011–2021, agriculture accounted for 4.3% of the World's GDP [1,2]. In 2020, statistics show that around 690 million people, or 8.4% of the world population, were starving [3]. By 2050, the World Bank has estimated that the population will increase to around 9 billion. If farmers fail to increase crop yield more people will continue to starve increasing the disparity and the percentage of people who do not have food. Farmers will have to increase food production by '70 percent' by the year 2050 [3, 4]. However, due to limited fertile and usable land, farmers will have to increase efficiency and production on the land that they already have. While most data shows that our current crop yield is increasing, it does not account for increasing land usage (the overall land area utilized for agricultural purposes) [3]. However, the availability of agricultural land is a finite resource [3]. This has led to the development of a new intricate field of technology, Climate Smart Agriculture (CSA) [3]. By taking advantage of this innovation and cutting-edge technology evolution in the area of agriculture, farmers can maximize efficiency and crop yields of their crops while keeping the same amount of land due to data that is usually not available or extremely timely and costly to obtain [3,5–7].

The principle of smart agriculture is to use data collected and provided by various systems to optimize the farming systems [7]. The main idea of smart agriculture is to solve the upcoming food crises we will face by increasing crop quantity and quality while also reducing the amount of resources, human labor, money, land, and so forth, poured into farming [3, 5]. Technologies 'range from IoT and robotics to drones and AI' [8]. The process of using most of these technologies is sensor-based and software-managed in such a way as to reduce prices, increase yield, increase quality, and improve customer experiences. Using these technologies allows for the

DOI: 10.1201/9781003676102-11

overall efficiency to be drastically increased and improved [6–8]. Smart agriculture has the potential to immensely impact the way food is grown in many areas of farming. One of the fields that smart agriculture will significantly impact is crop monitoring.

Before farmers started using this technology, this process would be arduous, timely, and costly. On top of all of this, the information was not always completely accurate or all the information needed was not present. This is due to human error and the usage of only human observation skills and primitive technology. However, this technology offers a solution using IoT technology and sensors. IoT technology allows farmers to track crop growth [4]. Sensors offer an increased aggregate of information ranging from crop and weed growth to weather and irrigation reports [9–11]. It helps eliminate many elements of surprise and danger that may be present in agriculture. Livestock is another field where its technology is applied. It gives farmers access to data on animals' health, food consumption, reproductive cycles, and nutrition levels [12, 13]. All of these result in a healthier flock and healthier food. Farming takes a long time. From planting the seeds to irrigating the crops to the final harvest, this all consumes an extensive amount of time. However, smart agriculture alleviates this challenge. With self-driving tractors, seeding drones, self-harvesting drones, and the like. farmers can devote less time and resources to the process [3]. These technologies will influence the field of agriculture in such a way as to provide better quantity and quality of food for everyone across the world while also lowering the resources needed to put into agriculture [3,6–7].

One of the many technologies that fall under smart farming is drones. Drones peaked in interest after scientists and farmers started to recognize the capabilities of drones. Drones, also known as Unmanned Air Vehicles (UAVs) or Unmanned Air Systems (UAS), have many applications within this industry. They provide the user with functions that otherwise would take the user hours and even days to complete. Some of these tasks include planting seeds, spraying pesticides and water, and harvesting crops [6]. In addition to these functions, drones also have sensors attached to them. These sensors give the user a way to detect water stress, dying plants, areas of concern in their fields, different types of plants and weeds [14], and the like. Most of these sensors use the electromagnetic spectrum to identify problems in plants [12, 13]. An example of this would be near-infrared sensors. These sensors use the near-infrared light emitted by plants during photosynthesis to tell if plants are dying [6]. Functions such as these give the farmer time to act on issues in the farm mid-season that they otherwise would not know about. People are not able to see other spectrums of light besides visible light. This is the reason why drone technology is essential in farming. Using drone photogrammetry, farmers can create a 3-D model of their field. While flying, drones capture images of the field. With these images, farmers can make topographical maps of the farmland in software such as Pix4DFields and Drone Deploy [7]. Different types of cameras on drones can also let farmers get different data. For example, RGB sensors can be used to check plant health [10,11,13,15]. The data that is acquired from these maps allows farmers to adjust the land for the crops to stay healthy. This data also allows farmers to see issues before they become too large and uncontrollable. This saves a lot of time and money for farmers [6, 7].

In addition to mitigating the work needed in the field, they also help in a multitude of ways when it comes to livestock farming. They help track the livestock and check where fences need to be fixed [5, 6, 12, 13]. When these drones have high-definition thermal imagers and night cameras, they can see other animals that are harassing herds of livestock. This tool is also used to help track poachers [6]. The benefit of using drones over doing the arduous tasks of the farm manually is the reduced time and cost savings. Drones are semi-automated which makes carrying out operations take reduced time and money on labor equipment. Drones also help identify problems beforehand which saves money and improves crop yield [3,5–7].

This chapter presents drones as an innovative and rising technology in the field of agriculture. It further discusses the types and configurations of drones, the basic architecture, various sensors, advantages and disadvantages associated with the various models/types of drones, their applications, and the challenges in their implementation in agriculture.

11.2 TYPES OF DRONES

Drones have many advantages and applications in farming. The two drones that comprise most of the smart agriculture market are fixed-wing and multi-rotor drones.

11.2.1 Fixed Wing Drones

It shows a typical design of a fixed wing drone. Its airplane-like design allows it to take off like an airplane, therefore, allowing it to be energy efficient and save battery [9, 16, 17]. Even though this is inefficient, this is the only time when the drone uses a heavy amount of battery life, allowing it to fly for longer and farther [18, 19]. When in the air, this type of drone flies by gliding through the air.

The wings, similar to an airplane's, are made to be aerodynamic and cut through the wind with minimal drag force to maximize the potential longevity of flight [20, 21]. Fixed-wing drones have the advantage over multi-rotor drones when it comes to flight time and area of coverage [9, 16, 17, 19]. Fixed-wing drones can fly for around 30 minutes to 1 hour 30 minutes. Their gliding flight style allows them to fly from 300 to 9600 acres [9, 11]. This makes them advantageous in large fields. Fixed-wing drones have more crash tolerance due to the high-durability foam inside of them. In the case of complete loss of power multi-rotor drones would lose flight, however with a fixed-wing drone they would still be able to glide and land safely [17, 18]. Using these drones comes with the main key advantage of increased battery life. The main challenge with using a multi-rotor drone is the constant drainage of battery life and the need to change the battery every 25 to 30 minutes. However, this challenge is almost completely negated while using a fixed-wing drone. Since it glides in the air, it does not require too much battery life [18]. This is what leads to the two main benefits of having a fixed-wing drone: increased area of coverage and increased time of flight. An overlooked feature is greater stability [18, 20]. Since they are built similarly to an airplane's design, they can withstand winds much better. On the other hand, these drones require wind planning ahead of time and are hard to manoeuvre. They cannot fly as precisely as a multi-rotor drone [18].

11.2.1.1 Models of Fixed-Wing Drones

Some models of fixed-wing drones are the SenseFly eBee X, Sentera PHX System, and the Parrot Disco Pro AG. The SenseFly eBee X is a drone that is used for mapping, It has a flight time of 90 minutes and travels around 500 acres per charge. The Sentera PHX System is used for crop analysis. This checks plant health, water stress, and so forth. The flight time is 59 minutes however; it covers an area of 700 acres. Finally, the Parrot Disco Pro AG is used for checking crop health. This drone has a coverage of 200 acres and flies for 30 minutes. All the different types of drones have different applications; hence farmers must know which ones to pick to benefit their farms. [7]

11.2.1.2 Multi-rotor Drones

A multi-rotor drone is a type of drone that has more than one rotor. Figure 11.2 shows a typical design of a multi-rotor drone. Typically, these drones tend to have four rotors, however, there are some other designs such as the hexarotor. The rotors allow this drone to have a full range of motion. By changing the speed of one of the motors, the drone can form an angle which then allows it to be in position to accelerate in a particular direction [22, 23].

Multi-rotor drones can only fly for 160 acres however they have many benefits. The main benefit that this type of drone has is that it is easy to maneuver around the field and fly close to crops [18, 24]. This allows it to collect data from sensors up close and take high-resolution pictures of the plant up close for data collection. Due to how light these drones are, attaching sensors and cameras is easier [18, 24]. The FAA limits the amount of weight on a drone therefore, it being light allows for more sensors and cameras on the drone. Multi-rotor drones are less expensive than fixed-wing drones [18, 24]. Their vertical takeoff and landing come in handy compared to the landing and takeoff of fixed-wing drones. Newton's Third Law states that every action has an equal and opposite reaction. This law is what creates the idea of the lift of a multi-rotor drone. The rotors of a multi-rotor drone create a downward force. This creates an opposite reaction that allows the drone to take off vertically. However, this lift force has to be larger than gravity, or else gravity will not allow the object to take off [22, 23]. This advantage however does not carry on to fixed-wing drones. Fixed-wing drones need a large area to land and take off, however, since multi-rotor drones take advantage of this law, they can take off vertically which takes up less space [18, 24].

11.2.1.3 Models of Multi-rotor Drones

There are many models of drones under both fixed-wing and multi-rotor. Some examples of multi-rotor drones are the DJI Phantom 4 Multispectral, DJI Agras T30, DJI Agras T20, and DJI Mavic 2 Pro. The DJI Phantom 4 Multispectral gathers information on plant health, nutrient levels, and soil moisture. In addition, it can also capture images in visible and near-infrared light. Another key feature that it possesses is the ability to spot plant stress and insect infestations [25]. These applications allow the farmer to gather data on their crops to find the best course of action to aid the crops' growth and increase crop yield. [7, 26].

Both the DJI Agras T30 and T20 are spraying drones. Both have their benefits over the other. The DJI Agras has a 30-liter tank. It has 16 nozzles for wide coverage and shoots around 8 liters a second with a 9-meter spraying range. This drone can cover around 40 acres in one hour and can fly during harsh weather. This allows farmers to irrigate their crops during heavy winds or unfavorable conditions. The DJI Agras T20, in addition to being a spraying drone, can also spread seeds, fertilizers, and other solid materials.This drone has 8 nozzles with a width of 7 meters in the spraying range. It can fly 12 hectares in one hour. In addition, it also has advanced obstacle avoidance technology and the ability to fly at any time of day and in harsh weather conditions. However, this drone does have some downsides. It has a max flight time of 15 minutes depending on the load. This could be annoying for the farmer to keep having to go back to the drone. Even though there is this downside, it is still overall a great drone since it is both a seeding and a spraying drone. [27]

Both fixed wing drones and multi-roto drones have their advantages and disadvantages. Table 11.1 discusses various properties associated with fixed-wing and multi-rotor drones.

Table 11.1 Fixed Wing versus Multi Rotor Drones Property Comparison

Properties	*Fixed Wing*	*Multi Rotor*
Distance	High	Low
Crash Tolerance	High	Low
Stability	High	Medium
Precision	Low	High
Manoeuvrability	Low	High
Cost Efficient	Low	High
Takeoff	Plane-like	Vertical

11.3 DRONES CLASSIFICATION BASED UPON THEIR APPLICATIONS

Under these two main types of drones, there are four others that each have their individual jobs but work in conjunction. These include sprayer drones, seeder drones, drones used to collect information, and drones used to map the area [3,5–8,12,9–11,13,15–24,28–30]. When these drones are in sync they work best. Drones used to collect information can provide the sprayer drones with information on which plants require more water or where pests are affecting the plants more. With this information, the sprayer drone can figure out where to spray more or less water, herbicides, and pesticides [30, 31].

11.3.1 Sprayer Drones

A key component of the agriculture process is irrigation and spraying pesticides, fertilizers, and herbicides. Figure 11.3 is a typical example of a sprayer drone. This, however, takes an extensive amount of time, money, and manpower and usually ends up not being completely accurate [3,5–7]. Sprayer drones are the solution to all of these issues. Using the data provided, farmers can plan out a flight pattern that will accurately target all the areas in need of these resources. Farmers can choose to irrigate some areas more or less depending on the information [24,28–31]. This will make sure that the resources are used to their maximum efficiency and will make sure to save the farmers time, money, and manpower [3,5–7].

11.3.2 Seeder Drones

Seeder drones are extremely useful to farmers. Figure-4 is a typical seeder drone. Much like sprayer drones, seeder drones rely on outside sources of information to calculate where to shoot seeds. For example, the seeder drones would use a topographical map of the area to figure out where crops would grow most successfully and where it would be an extremely harsh environment where plants would not grow [32]. These drones would not shoot seeds in rocky areas because they knew that crops would not be able to grow. According to Emily Nonko, drones usually fly '130 to 200 feet in the air' and deploy around 'five pods per second at a strong enough velocity that they embed into the soil' [33]. Many startups have designed seeder drones that have an uptake rate of 75% and lower costs by 85% [32]. This helps farmers profit much more from the same amount of land which is the main goal of smart agriculture. In addition to applications in agriculture, these drones have applications outside of agriculture as well. They are used in missions to replant forests to reverse the effects of deforestation [33].

11.3.3 Information Collecting Drones

Drones that collect information are crucial for farmers. They are what make smart agriculture smart agriculture. These drones provide the farmer with information such as plant health, water stress, disease, identification of weeds, and so forth. All of these factors come in handy in terms of deciding how to help solve issues in plants. If farmers notice a trend in the data, for example, if a certain area is always dehydrated, they could use this information to solve these problems. This saves time and money which is a theme seen throughout all smart agriculture [3,6–7]. Drones that are used for mapping can be useful for planning out where each crop would go and figuring out areas of irrigation, and the like. These maps can also work in conjunction with data gathered from other drones to make a map of where plants are more exposed to a particular pathogen, where plants are dehydrated, and so forth [34–36]. All four of these drones work towards a common goal of producing more yield while lowering the manpower, time spent, and money.

11.3.4 Sensors in Drones

Drones can carry out specific tasks with the help of a wide variety of sensors. These sensors can range in application from basic movement capability to advanced monitoring. FIgure 11.1 discusses various sensors and their high-level classification.

Table 11.2 summarizes various applications for the types of sensors and their classifications.

11.3.4.1 Movement Based Sensors

Sensors are an essential part of smart agriculture. They provide data on plant health, weather conditions, water stress, pathogens, and the like. Most of the sensors used to gather data are ones focused on looking into the electromagnetic spectrum. Humans can only see the visible light part of the spectrum, but the rest cannot be seen. Therefore, humans cannot collect

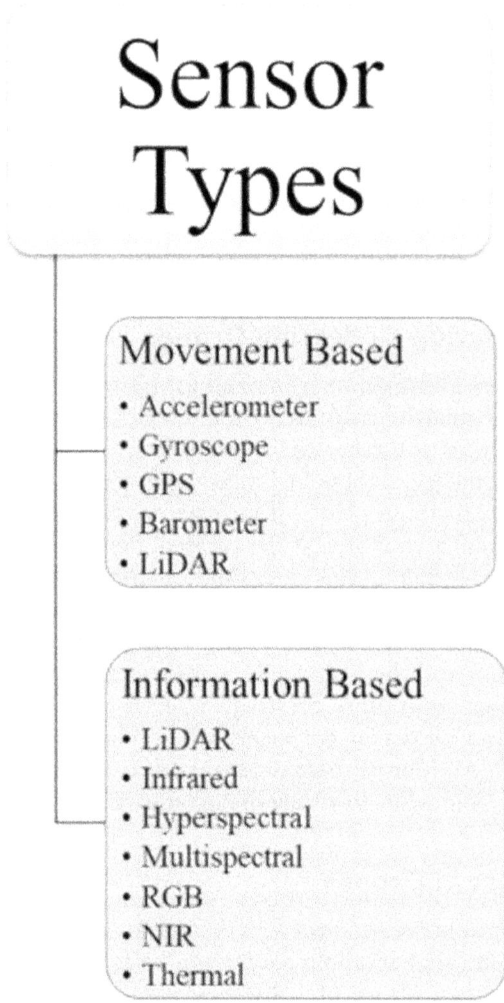

Figure 11.1 Classification of Sensors.

information as effectively as drones can. However, there are many sensors besides ones that view the gathered information on the plants, livestock, and crops. These include the barometer, GPS sensor, LiDAR sensor, 3-axis accelerometer, and 3-axis gyroscope. They contribute to the base function of the drone, driving, obstacle detection, and manoeuvring.

Barometer: The barometer is a sensor that detects the height of the drone and is also used to control the height. The property that allows

Table 11.2 Sensors in Drones

Classification	*Type of Sensor*	*Application(s)*
Movement based	Barometer	Detects height/controls height
Movement based	GPS sensor	Used for making routes for drones/stops drift of drone
Movement based & information based	LiDAR sensor	Terrain mapping/helps avoid obstacles in autonomous flight **Problem:** Expensive/not good quality
Movement based	3-Axis accelerometer	Detects position and orientation/controls movement of drone/provides acceleration in all 3 axes
Movement based	3-Axis gyroscope	Provides angular motion and acceleration in all 3 axes/more accurately calculates angles
Information based	RGB sensor	Detect plant health/mapping and scouting/ plays a role in machine learning **Problem:** Not accurate plant health data
Information based	NIR sensor	Detects plant health and stress
Information based	Thermal sensor	Detects water stress and diseases in livestock/ can be used at night to find predators to livestock
Information based	Multispectral sensor	Has 3–5 bands (red, green, blue, red edge, and NIR) which farmers can choose from/ detects chlorophyll content. Pairs well with DLS (Downwelling Light Sensor) because multispectral data is affected by sunlight and DLS removes that factor.
Information based	Hyperspectral sensor	Detects soil quality, water stress, and early signs of pathogens/differentiates and can tell type of plant based on light emitted [62] **Problem:** Heavy/expensive/hard to stabilize

this to happen is the change in atmospheric pressure the higher something goes. Even though this statistic is already detectable via the GPS sensor, it is inaccurate and is crucial to have an accurate reading of the altitude. This enables it to stabilize the drone for videography and for flying precisely [37].

GPS sensor: The GPS sensor is crucial to a drone. By connecting to a system of orbiting networks the sensor allows the drone to know its location and position. This is extremely useful in autonomous tasks as it allows farmers to plan out flight patterns for the drones to stay on. This GPS sensor is used for making a route for the drone to fly on to complete a task. These help a lot when the pilot is not necessarily driving the drone. These help the drone stay on course otherwise any form of drift could move them off the path which could waste time and could potentially harm crops or break the drone.

LiDAR sensor: The LiDAR sensor is a sensor that is used for terrain mapping, autonomous flight, and avoiding obstacles. The LiDAR sensor shoots out pulsed light waves which bounce off the surrounding objects and inevitably return to the sensor. The sensor then uses the time each light wave used to return to calculate the distance of objects and obstacles in the area. This enables drones to make a map of the surrounding landscape. This can be applied to seeding and planning autonomous flight patterns. Despite the various benefits provided by this drone, there is a major disadvantage that this sensor has. This sensor is not common on drones due to its increased cost compared to other sensors. Despite its versatility, most farmers refuse to buy this due to how expensive it is.

3-Axis accelerometer: The 3-Axis accelerometer's main purpose is to measure the orientation and position of the drone. It does this by measuring the acceleration of gravity. This sensor detects position and orientation. It is crucial in making the drone stable in all three axes (X, Y, Z). It also provides the drone with acceleration in all three of the axes. This sensor is used alongside a controller to control the movement of a drone. It provides more or less electricity in certain rotors depending on how the pilot moves the joystick. This is crucial for the movement of drones [38].

3-Axis Gyroscope: Finally, the 3-Axis Gyroscope is used to provide angular motion and acceleration in all 3 axes. This also provides more stability to the drone and has higher accuracy in programs. The difference between this sensor and the 3-Axis accelerometer is the 3-Axis accelerometer will not calculate the angular velocity and it will also not take it as input however the 3-Axis gyroscope will. This sensor is crucial for drones set on a certain flight path and for autonomous drones.

These sensors are the foundation for the basic movement and maneuverability of the drone. They also open the door for autonomous flight patterns with the aid of sensors such as LiDAR, GPS, barometer, and the like. Despite providing the base, these sensors do not provide the farmer with any data about plant health, weed management [39], cattle herd management, and so forth.

11.3.4.2 Information-Based Sensors

Many other sensors detect plant and crop health. These sensors are what truly make smart agriculture. They differentiate a person flying a drone from smart agriculture. All of these sensors rely on some form of light besides visible light to collect data. The data that these sensors collects helps farmers to make educated decisions to solve problems that are occurring in their fields and even increase yield with the same amount of land. Some sensors that collect data include the RGB sensor, NIR sensor, thermal sensor, multispectral sensor, and hyperspectral sensor.

RGB sensor: The RGB sensor has many applications. This sensor allows the farmer to check plant health using VARI (Visible Atmospherically Resistant Index). This algorithm uses color correction to reduce reflectance, scattering, and other atmospheric effects that can mess up the calculations for plant health [40]. Essentially, what it does is exaggerate the color green to check plant health. However, this is not to be used instead of the NDVI (Normalized Difference Vegetation Index). RGB sensors have the benefit of mapping and scouting abilities. Using traditional ways, farmers would have to walk around and use their tractors to go through the whole field and inspect the crops. This would take time and labor. However, with this sensor, it is extremely easy to view the whole field. It makes a map using aerial imaging and then converts it into an orthomosaic map. This is also useful because when farmers use tractors they could harm crops, however, farmers can inspect the crops without even having to go outside now. One of the largest benefits of this sensor is that the data is simple to understand. This is because the data produced by these cameras is almost identical to visible light making it visible to humans. RGB cameras play a significant role in machine learning. It helps with object recognition, phenology, pathologies, and similar purposes [5, 7, 11, 40, 41].

NIR sensor: To do photosynthesis, plants need to capture visible light. Plants have evolved to reflect NIR light because it does not carry enough energy for photosynthesis, but they do bring lots of heat. As the plant starts to die, this mechanism slowly breaks down. As a result, it is not able to reflect NIR light. Near-infrared sensors take advantage of this by monitoring the difference between visible light reflectance and NIR reflectance. This calculation is known as NDVI or Normalized Difference Vegetation Index. A strong NDVI signal indicates healthy plants and the opposite for a weak NDVI signal. This becomes useful in locating problem areas in the crops. This can help pinpoint where to add more fertilizer and can also indicate the presence of weeds, pests, water damage, and the like. The application is extremely similar to the one shown in RGB however it is much more accurate. The name of the calculation hints at the reason why NIR sensors are much more effective than RGB sensors at detecting plant health. Normalized is the keyword that is NDVI but not in VARI. The fact that it is a normalized calculation means no matter how cloudy or sunny it gets the data produced by the sensor will be the same, however the same cannot be said for the VARI calculation. To put this in simpler terms, VARI can vary.

This is why it is essential to have NIR sensors checking the data on plant health. Another key benefit that NIR sensors have is the time it takes to detect issues. Plant stress shows up in NIR quicker than in visible light. Sensors such as RGB use visible light to detect plant stress, however, this takes a long time. NIR can usually find issues faster than people can with their naked eye. This extra time can maybe allow farmers to fix the issue before it becomes detrimental and harms a large chunk of the crop yield [13,15–17].

Thermal sensor: Thermal sensors detect the radiated temperature of an object. The latest models in this field of sensors are light enough to be carried by a drone. A thermal sensor can help tell whether a plant is getting enough water or not. When a plant is under water stress, it lowers the rate of evapotranspiration. When this gets lowered, the canopy temperature gets increased. Using this information, the sensor detects which plants are cooler and which ones are warmer. However, there are some flaws with this. Other factors can affect the data. This includes wind, sun exposure, breezes, and so forth. Therefore, scientists need to research more into this sensor. Thermal cameras are also extremely useful for livestock. Livestock are affected by many diseases such as mastitis. Usually, this disease is detected in the milking process, however since most of this is automated there is no way to tell without first testing it in a lab. Thermal sensors provide a solution to this. In the past, farmers were able to tell because the cow's udder would rise in temperature. However, using thermal sensors, farmers can now detect this disease and stop it from becoming a problem. These sensors could also be used to stop predators from harming livestock. At night, farmers cannot see their surroundings. Using a thermal sensor, they can detect predators and stop them before they harm the livestock [11,15–17].

Multispectral sensor: Using multispectral sensors, farmers get access to an aggregate of data. This sensor almost always detects light in three to five different spectral bands. For example, a sensor could detect green, red, and near-infrared while another detects blue, green, red, near-infrared, and red edge. Being able to mix and match the different bands of light allows farmers to have one sensor detecting different information about crop health. For example, this sensor is given the same functionality as NIR sensors. This means it can produce the same data and use the same vegetation indices, NDVI. Red edge is an extremely important part of this sensor. Red edge is a spectral band located between NIR and red. Plants have a high level of reflectance in this band, and this usually correlates with plant health. Red Edge also has vegetation indices called Normalized Difference Red Edge (NDRE). Both NDVI and

NDRE are extremely useful, however, there are certain times when one outshines the other. From early to mid-growth stages, NDVI is much more effective at portraying differences in chlorophyll content and canopy density. However, from mid to late season the NDRE is much more effective at indicating chlorophyll content and total biomass. There is one major issue with the multispectral sensor. Sunlight is a factor that can change the data. The sensor could calculate the sunlight's reflectance into the data resulting in a variable. This stops the sensor from being able to collect data over time. However, there is a way to fix this by pairing this sensor with a Downwelling Light Sensor (DLS). A DLS is a sensor that detects the amount of sunlight in each of the spectral bands. This allows for the data to be compared to the sunlight of that day allowing the multispectral sensor to be accurate for data in a long period [7, 10, 11, 16].

Hyperspectral sensor: Hyperspectral imaging helped revolutionize imaging by getting a multitude of images at various wavelengths. This allows it to gather better spatial and spectral information when compared to other methods such as multispectral imaging, spectroscopy, and RGB imagery. Hyperspectral sensors have many key features that can help farmers in the early stages of their farming. These sensors detect soil quality, water stress, and early symptoms of pathogens and disease. These key features can help plan out where crops should be placed for optimal growth and can help treat plants that require the right treatment. Hyperspectral sensors can identify a specific type of plant by measuring the color of the light it reflects. This could be useful to farmers for picking out herbicide-resistant weeds. There are many issues with this sensor. It is bulky and heavy which means that the actual drone has to be lightweight, or it will not be allowed to fly under the FAA guidelines for drones. This heavyweight and large set of features results in it being expensive for farmers to buy. This makes this sensor a heavy investment. The stability of this camera is another issue. Due to its size, it is much harder to stabilize this camera on smaller drones. Another major problem with this camera is calibrating it to work on a low-flying drone where light conditions vary across the field. Even though this sensor is plagued with problems its benefits make it a worthwhile investment to help the progress of the farm [41, 42].

11.4 AGRICULTURAL APPLICATIONS OF DRONES

Drones are a breakthrough in agriculture and benefit this industry in many ways. They can reduce the time for things that once took days to only taking hours. With the use of various kinds of sensors, drones have a multitude of applications in smart agriculture. From the early stage of farming to harvesting the crops the use of drones is always needed. These drones can aid the farmer in ways such as land imaging, crop health monitoring, irrigation monitoring, weed identification, cattle herd monitoring, damage assessment, soil analysis, planting, and spraying. Within this range of applications, there are two broad categories: monitoring-based applications, and actionable applications. Monitoring-based application mainly allows the area under supervision to be scanned, monitored and reported for collecting the data for reporting requirements. On the other hand, actionable applications would allow a drone to directly take control and perform some set tasks within a specified area of interest [43–48].

11.4.1 Monitoring Applications

The following are the general monitoring applications that could be deployed using drone technology for smart agriculture.

Land imaging: In the early stage of farming, land imaging is extremely useful. Before farmers were able to use drones for this, they had to use satellite imagery. However, this resulted in there being two issues. This was expensive and the resolution of the image was terrible. Drones solved both of these issues. Two main sensors help with this application. Those are RGB and LiDAR sensors. How it works is the drones take multiple images of the area of interest and then stitch them together through a process of photogrammetry. This image is called an orthomosaic map. After this, the image is georeferenced and made to scale. This application is used for planning out the field. Using data collected by other sensors, farmers can tell where the most optimal spot is to grow their crops and increase the yield. Planning out where crops should go with data from previous years is also important. Farmers could use previous years' maps to see where crops were struggling to grow and where they were thriving. Using information from past years can help them plan out their farm to have optimal growth. This minimizes the risk of crop damage from factors such as weeds, pests, and the like. It also allows them to analyze their irrigation system and see which crops get the most water and which ones are under water stress. Using thermal cameras, they can upload this data onto the map, and using a color scale, they can tell where they need to irrigate plants more or less. This same concept applies to almost any major issue that occurs on the farm. A farmer could easily input the data into the map and see where crops get harmed most and where they are growing best [43–46].

Crop health monitoring: Every year, farmers lose 10%–20% of crops due to health-related problems, or $100 billion to $200 billion. This makes crop health monitoring one of the most important applications of drone technology. This application is made possible by many sensors such as NIR, RGB, multispectral, and so forth. This application allows farmers to check on the health of their crops without even needing to go outside. This step would usually take hours for farmers to complete on a large farm and would require a large amount of labor. An application that once took days to complete now only takes a couple of hours with the aid of drone technology. In conjunction with land imaging, farmers can pinpoint exactly where the issue is and can stop it before it becomes a major issue. This also helps in saving money. Farmers usually do not know how much fertilizer is needed for crops or how much water is needed. Thanks to this application, farmers will know that one part of the farm is doing just fine, however, the other part of the farm is struggling to survive. Using this information, a farmer can adjust irrigation and fertilization, pouring in more resources where they are needed and putting in less where they are not. This application aligns with smart agriculture's three core principles: saving money, time, and increasing yield and quality [42–46].

Irrigation monitoring: Irrigation is a huge part of agriculture. This results in the monitoring of what is crucial to plant health. Using spectroscopy and thermography, which is done by infrared and thermal sensors, farmers can reveal crop wilting. It can also be used to detect leaks in equipment and irrigation channels. They can also use airborne laser scanning technology, otherwise known as land imaging, to understand the topography of their land. They can also use the technology to find catchments that reveal the direction of water flow at the bottom of each tree and can be used to find land features that may affect crop health and soil erosion. Therefore, farmers can know which crops to irrigate more and which ones to irrigate less [45–47].

Weed identification: Weeds are a major problem on any farm. They can hinder the growth of crops and even kill them. This in turn reduces the yield by a significant amount. In total 10% of crops are lost to weeds across the world. To put in perspective how much that is, around 200 million metric tons of grains are lost. Weeds account for 100 million metric tons of loss. Fortunately, with the usage of sensors, farmers can now prevent this issue before it gets out of hand and makes a detrimental impact on yield. Sensors and drones are an efficient and environmentally safe way to control the weed population. It enables precise and continuous monitoring and mapping of weeds and their infestations. It optimizes weed treatments and makes it more economically efficient. This system helps reduce herbicides. This in turn makes the food quality better because it is healthier due to it having fewer herbicides [42–48].

Cattle herd monitoring: Livestock has a significant effect on our food supply. From meat products to dairy products, none of those would be possible without livestock. However, there are still some problems with livestock. Drones work to solve this issue. Cattle herd monitoring makes keeping livestock at farms much easier. Drones are used in many ways when it comes to livestock farming. They help track the livestock and check where fences need to be fixed. When these drones have a high-definition thermal imager and night camera they can see other animals that are harassing herds of animals. This tool is also used to help track poachers. Drones could easily keep an eye on the livestock and could analyze their behavior. This is crucial because it could reveal injuries to the animal or diseases. This could stop an injury or disease before it gets out of hand and kills or seriously harms the animal [49].

Damage assessment: Natural disasters are bound to affect any farm. However, depending on how large a farm is the workload of finding the affected area can be severely large. From 2008 to 2018 $49 billion was lost in Asia, $30 billion was lost in North Africa and sub-Sahara, and $29 billion was lost in the Caribbean and Latin America. To effectively measure the disaster area agricultural insurance companies, conduct an agricultural insurance disaster loss survey and apply drones to scan the area. Drones are equipped with the abilities of quick maneuverability, high-resolution imagery, precise-positioning data acquisition, accommodating other devices, and a convenient maintenance system. These features enable drones to be more efficient and accurate while scanning affected areas. Drones help save a lot of manpower and resources and produce an effect that is more efficient and accurate. In turn, drones have become significantly more useful in the agriculture insurance and assessment sector. Not only is this useful for insurance, but also useful for farmers. This technology can detect crop damage from sources besides natural disasters. This allows the farmers to see where and how the problem is occurring and then work to fix it. This mitigates future crop loss [50, 51].

Soil analysis: Soil analysis is crucial for any farm. Even if the climate is good for farming it would not matter because without healthy soil, farming will not be successful. Before the use of smart agriculture, farmers would have to send a sample of their soil to a lab where they would test it. After they study the sample and generate a report with data such as nitrate levels, potassium, phosphates, soil pH, and organic matter. With this information, farmers would know how to tweak their nutrient levels using fertilizers and soil conditioners. This takes a long time and is not completely accurate. Fortunately, with the use of multispectral sensors, farmers can do this in real-time. There would be a flyover that would provide the farmer with data on the whole field and not just represented area samples. This is useful because if the soil needs different nutrients than another area, but a farmer represents the whole area with one sample the plants will wilt. This application increases yield while saving time [43–46].

11.4.2 Actionable Application

The following are the general actionable applications that could be deployed using drone technology for smart agriculture.

Seeding: Planting crops is one of the most important steps. It is the whole basis of the farm, and it is what results in yield. However, this process takes a long time and is arduous. It can take days for a farmer to finish planting a whole field even with a substantial amount of labor. This is where drones come in handy. Some drones plant crops and are called seeder drones. These systems shoot pods with seeds and plant nutrients into the soil, providing the plants with all the nutrients necessary to maintain life. This allows them to save time and manpower. They are also used outside of agriculture in ways such as reforestation. The drones fire at a rate of one seed a second or 100,000 a day accounting for battery life, charging, and reloading seeds. Therefore, if there were 60 teams of drones, every year 1 billion trees could be planted [32, 33].

Spraying: Sprayer drones increase crop yields by almost 5%. Spraying, like planting, takes a long time. Manually it can take a day or even more. However, drones work to solve this. Many drones can irrigate a whole farm in just a couple of hours. All that is required is someone to watch the drone and using autonomous fight patterns they do not even need to maneuver the drone [24,28–31].

Harvesting for Fruits: Around $30 billion worth of fruit is lost due to it rotting on trees. Currently, there is a shortage of fruit pickers, and the shortage is only expected to grow [52]. By 2050, it has been predicted that there will be 5 million fewer fruit pickers. This causes an increase in fruit not being picked which results eventually in more rotting fruit leading to an increase in loss. Picking fruit on time is crucial to the amount of output. If fruit is even picked 2 weeks late almost 80% of it is wasted. Around 10% of fruit is wasted worldwide due to this major issue. This is the same amount of fruit that the whole European Union consumes every year. This becomes a major issue for farmers. However, drones provide a solution to this current challenge. An Israeli startup, Tevel Aerobotics Technologies, has made a drone that picks ripe fruits off trees. This solution provides a more efficient way to pick fruit. Because this is a robot it can work 24/7. It does not need to sleep, take a break, or do other activities that humans are required to do. This drone is tethered to the charger so that it never has to stop to be charged. This new groundbreaking system allows farmers to increase yield while also saving money. Farmers no longer must pay the salary of all the hundreds or even thousands of workers that they had before but now only need to pay an initial investment which saves a large sum of money long term [46–48].

11.4.3 Challenges with Drone Technology for their Use in the Field of Agriculture

Most of the challenges presented in drone technology boil down to three categories: money, time, and safety. In the end, drone technology is extremely useful, however has many current challenges that are still being worked out. People who develop this technology are working to lower prices and increase safety measures to open up new applications for drones and make them more financially suitable for farmers. Some of the major challenges include FAA safety guidelines, complicated systems for farmers, slow data processing, high initial investments, battery drainage, internet connection, and weather-related challenges [53–56].

11.5 FEDERAL AVIATION ADMINISTRATION GUIDELINES

The Federal Aviation Administration, also known as the FAA, enforces regulations on manufacturing, operating, and maintaining aircraft. This federal organization also limits drone usage and creates rules. For example, only drones between 0.55 and 55 pounds can fly in the air. However, these drones cannot be used for any function that the owner wants. There is a list of specific flight operations a drone can do. These are crop monitoring and inspection, research purposes, development, educational uses, power line, and pipeline inspection in hilly or mountainous terrain, antenna inspections, aiding rescue operations, bridge inspections, aerial photography, and wildlife nesting area evaluations. On top of this, they cannot be flown in a heavily crowded area. In addition to rules for drone operations, there are pilot licensing rules. All drone pilots are required to have a license. People can obtain this by using a private pilot license or UAS pilot license. It is required for the Pilot in Control (PIC) to have this license. The PIC, contrary to popular belief, does not have to be the person flying the drone however the PIC has to be present while flying the drone to oversee and monitor flight operations. There are many flight rules that need to be followed when flying. These are the max altitude is 400 ft, and the range is only as far as the PIC can see with the naked eye, drones should be flown within the specification of the operator's manual, they cannot be legally modified to carry items, they cannot be flown over crowds and populated areas, the PIC needs to consider weather conditions and obstructions before takeoff, and flights are only allowed in between 30 mins before sunrise and 30 mins after twilight. All these rules limit the potential of drone technology in smart agriculture however allow the general population to remain safe while using this advanced yet potentially dangerous technology. This is one of the many challenges that drones face in smart agriculture. Just like these guidelines, many other countries across the world have guidelines. However, most of the guidelines boil down to the same principles with minor differences here and there. The main point of these is to keep the general public safe from this technology which can limit the potential output that drones can have [56–60].

11.5.1 System Complexity and Its Training

Drone technology is complicated. From the flying of the drone to the processing of the image, it is not easy to understand everything. Even though farmers learn how to fly the drone to get the license, there are still other systems that go alongside the base drone such as image processing, data collection, and so forth. This results in the need for them to be trained or hire someone to do all the complicated technological systems. This adds to the farmer's initial investment which could be upwards of $25,000. Reading the manual and understanding the functionalities of drones takes a large amount of time. So essentially, farmers have to choose whether to invest money and learn it quickly or invest in time and have slower beginning growth however save money in the initial investment [53–56].

11.6 DATA PROCESSING

Drones' systems will produce a large amount of complex data which will mainly be comprised of images and videos. The problem with data processing in drone technology is it takes data for hours or even days each time new data is loaded. This, in turn, halts productivity and requires there to be someone operating the computer to monitor progress. With this large time gap, the farmer's effort to find out issues ahead of time could go to waste due to this. This challenge almost completely negates one of the pillars of smart agriculture from drone technology.

However, there is a way to overcome this challenge. Processing can also be done using a purpose-built, web-based service. Using this, the software runs autonomously and uploads images to the cloud. Their images are processed rapidly and are given back to users. This, however, also adds to the initial investment which can seem daunting to farmers [37,57–60,61].

11.6.1 System Affordability

The initial investment in drone technology is daunting to many farmers. Drones alone cost anywhere from $1500 to $25,000. But farmers cannot just buy the drone. On top of this, they have to buy extra accessories to function such as internet, controllers, mobile electronic parts, and closed-circuit TV cameras. This almost doubles the initial cost of the drone. Not to mention that they also must buy a license to fly the drone. Even though farmers do eventually get back their money from the benefits of drone technology, the initial investment is daunting which steers many farmers away from the technology [59–61].

11.6.2 Power Issues with Drones

Battery is a major challenge for drones. The average multirotor drone can only fly six acres in 25 minutes per charge. This means for surveying fields, planting, spraying, or data collection a farmer would need to change the battery 12 to 15 times. And that number is the average field size. For large farms, this hindrance is a nightmare. The cost of batteries is high and having to buy upwards of 12 per drone active will substantially increase the initial investment. Not only does it add to the cost, but it also drains efficiency. Farmers have to run out into the field and resupply a drone every 30 minutes, which is a tedious task. This harms efficiency rates and wastes time by diverting the farmers' attention. This hinders long flight patterns for data collection, spraying, and so forth. Not only does it have a short battery life it also takes a long time to charge. According to Future Farming, it takes around 60 to 90 minutes to charge a battery.

However, companies are finding ways to overcome this obstacle for agriculture drone technology. A company called StoreDot had managed to completely charge a battery in just 5 minutes. This is huge progress toward overcoming this challenge [37,56–61]

11.6.3 Network Connectivity

Most farms are in remote areas which are a long away from cities. This makes for a lack of internet connection, in around 18% of farms, which is necessary for data processing, planning fight patterns for drones, seed entries, and so forth. To get this internet connection, farmers need to pay an internet provider for access which increases the initial investment. Farmers could also 'buy a drone capable of capturing and storing data locally in a format that can be later processed' (Agrotech). This only solves the challenge of data processing and also adds to the initial cost [54–57].

11.6.4 Weather-Related Challenges

In any flight-related function for drones, extreme weather conditions, such as rain, thunder, hail, and so forth, pose a major challenge. Unlike planes, the weather always messes up the flight pattern and poses a major threat to drones. If a drone is flown through a storm, it could get lost or severely damaged. The heavy rains and high wind velocity can mess with the signals in the drone which cause some of the sensors to stop working or even crash. This could completely ruin the image quality if the drone or the camera on the drone gets damaged [53–56].

11.6.4.1 Potential Economic Impact of Drones on Agriculture

The market for agricultural drone technology is expected to skyrocket in the next couple of years. It is estimated that by 2032 the market will surpass an evaluation of US$11.08 billion. These precision farming systems can increase yields by 5% [24]. This also opens up new jobs for people around the globe. Jobs such as maintenance, pilots, software managers, and the like. are opened up because of this industry. In India alone, this new technology has opened up 500,000 jobs and has opened up an economic opportunity of $25 billion due to the 15% increase in productivity and production rates. With all these factors at play, it has managed to boost the GDP of India by 1.5% [2]. Across the world, this technology is enabling a 5% productivity increase which would allow more crops and better crops to be put into supermarkets. This would lead to more produce being available to buy and decrease world hunger significantly.

Around 11% of the total drone market is the agricultural drone market. Drone companies are expected to draw in around $50 billion in investments. These systems are heavily reducing issues such as world hunger which is plaguing our modern world while also opening up several potential job opportunities and increasing productivity, efficiency, and total production [— 61-62].

11.7 CONCLUSION

This chapter has presented a detailed discussion about the usage of drones in the field of smart agriculture. Innovative technologies must be explored and adopted to enhance agricultural productivity. Drones are one such technology that can aid farmers in various applications as discussed in this paper. There are challenges associated with the use of drones. Those challenges are likely to be addressed as the field progresses with time and new advancements in hardware and software. The economic landscape for agriculture is likely to see a dramatic increase due to the use of this technology. It opens up new potential for the field.

REFERENCES

1. A. O'Neill, "Share of Economic Sectors in the Global Gross Domestic Product (GDP) from 2011 to 2021", *Statista*. www.statista.com/statistics/256563/share-of-economic-sectors-in-the-global-gross-domestic-product/. Accessed 19 July 2023.
2. "Drones Can Boost India's GDP by up to 1.5%, Address Challenges for the Country's Key Agriculture Sector", *The Economic Times,* 18 Oct. 2022. https://economictimes.indiatimes.com/news/economy/agriculture/drones-can-boost-indias-gdp-by-up-to-1-5-address
3. P. Dev, S. Khandelwal, and S. C. Yadav, "Climate Based Smart Agriculture: Need for Food Security and Sustainability", *International Journal of Environmental Climate Change*, vol. 13, no. 3, pp. 224–231, 2023.
4. S. Herrick, "RGB Versus NIR: Which Sensor Is Better for Measuring Crop Health?" *Botlink*, 17 Aug. 2017, https://botlink.com/blog/rgb-versus-nir-which-sensor-is-better-for-measuring-crop-health.
5. A. Rejeb, K. Abdollahi, H. Rejeb, and Treiblmaier, "Drones in Agriculture: A Review and Bibliometric Analysis", *Elsevier Journal of Computers and Electronics in Agriculture*, vol. 198, pp. 1–19, 2022.
6. Q. Ren, R. Zhang, W. Cai, X. Sun, and L. Cao, "Application and Development of New Drones in Agriculture", *IOP Conference Series: Earth and Environmental Science*, vol 7, 440, pp. 1–3, 2020.
7. M. Canicattì, and M. Vallone, "Drones in Vegetable Crops: A Systematic Literature Review", *Smart Agricultural Technology*, p. 100396, 2024.
8. D. Casto, "Precision Farming vs. Digital Farming vs. Smart Farming: What's The Difference?" *DTN*, 2021, www.dtn.com/precision-farming-vs-digital-farming-vs-smart-farming-whats-the-difference/
9. Torres-Rua, "Drones in Agriculture: An Overview of Current Capabilities and Future Directions", *Utah Water Research Laboratory*, pp. 1–9, 2017. https://conference.usu.edu/uwuw/includes/AlfonsoTorres_DronesinAgriculture.pdf?v=1.22
10. M. Ungerer, A. Bowmaker-Falconer, C. Oosthuizen, V. Phehane, and A. Strever, *The Future of the Western Cape Agricultural Sector in the Context of the Fourth Industrial Revolution (Synthesis report)*, Western Cape Department of Agriculture (WCDoA), together with the University of Stellenbosch Business School (USB), 2018.
11. P. Daponte, "A Review on the Use of Drones for Precision Agriculture", *IOP Conference Series: Earth and Environmental Science*, vol. 275, pp. 1–10, 2019.
12. Rangone, "Innovative Agribusiness. Which strategy for a dynamic development?", *Agribusiness and Information Management*, vol. 11, no. 2, pp. 5–15, 2019.

13. R. B. Kalamkar, M. C. Ahire, P. A. Ghadge, S. A. Dhenge, and M. S. Anarase, "Drone and its Applications in Agriculture", *International Journal of Current Microbiology and Applied Sciences*, vol. 9, pp. 3022–3026, 2020.
14. B. S. Chauhan, "Grand Challenges in Weed Management", *Frontiers in Agronomy*, vol. 1, pp 1-4, 2020.
15. N. Rani, K. Sinha, A. Chaudhary, and M. Mohanty, "Drone: The Green Technology For Future Agriculture", *Harit Dhara*, vol. 2, no. 1, pp. 3–6, 2019.
16. L. J. Thompson, Y. Shi, and R. B. Ferguson, "Getting Started with Drones in Agriculture", *Nebraska Extension*, pp. 1–12, 2017. https://extensionpubs.unl.edu/publication/g2296/na/html/view
17. T. McKinnon, "Agricultural Drones: What Farmers Need to Know", *Agrobiotix*, pp. 1–9, 2016.
18. J. Beth, "Fixed Wing vs Multirotor Drones for Surveying", *Coptrz*, 24 Feb. 2021, https://coptrz.com/blog/fixed-wing-vs-multirotor-drones-for-surveying/.
19. D. Reed, "Multi-Rotor vs. Fixed Wing UAV Platforms: Considerations for Evaluating Capabilities and Limitations", *GEO Jobe*, 1 Dec. 2021, https://geo-jobe.com/drones-uav/multi-rotor-vs-fixed-wing-uav-platforms-considerations-for-evaluating-capabilities- and-limitations/.
20. "Fixed-Wing Drones: The Evolution of the Technology", *Wingtra*, 12 Jan. 2021, https://wingtra.com/fixed-wing-drones-the-evolution/. b. "Types Of Fixed-Wing Drones/UAVs Explained (+Pictures) – The Corona Wire". www.thecoronawire.com/types-fixed-wing-uavs-what-are-how-they-work/. Accessed 19 July 2023.
21. Allain, Rhett, "How Do Drones Fly? Physics, of Course!" *Wired*. www.wired.com/2017/05/the-physics-of-drones/. Accessed 19 July 2023.
22. V. Beliautsou, A. Fedorova, M. Syravatnikau, and V. Mokshin, "Development of a Stabilizing System for Quadcopter with Variable Thrust Vectors", in *IEEE International Conference on Information Technology and Nanotechnology (ITNT)* (pp. 1–5), May 2020.
23. Market.Us, "Agricultural Drones Market Is Anticipated to Surpass a Market Value of US$ 11.08 Billion by the End of 2032", *GlobeNewswire News Room*, 13 Apr. 2023, www.globenewswire.com/en/news-release/2023/04/13/2646682/0/en/Agricultural-Drones-Market-Is-Anticip ated-to-Surpass-a-Market-Value-of-US-11-08-Billion-by-the-End-of-2032.html
24. K. Bhawan, "Standard Operating Procedure (SOP) for use of Drone Application with Pesticides, for Crop Protection and for Spraying Soil & Crop Nutrients in Agricultural, Forestry, Non-Cropped Areas, etc.", Ministry of Agriculture, *Government of India*, pp. 5–24, Unknown.
25. C. A. Deutsch, et al., "Increase in Crop Losses to Insect Pests in a Warming Climate", *Science*, vol. 361, no. 6405, Aug. 2018, pp. 916–19, doi: 10.1126/science.aat3466.
26. "DJI – Official Website", *DJI Official*, www.dji.com. Accessed 20 July 2023.
27. M. Esposito, M. Crimaldi, V. Cirillo, F. Sarghini, and A. Maggio, "Drone and Sensor Technology for Sustainable Weed Management: A Review",

Chemical and Biological Technologies in Agriculture, pp. 1–11, 18 (2021). https://doi.org/10.1186/s40538-021-00217-8
28. E Ozkan, *Drones for Spraying Pesticides—Opportunities and Challenges*. https://ohioline.osu.edu/factsheet/fabe-540. Accessed 19 July 2023.
29. C. S. Kinnard, *CAES Investigates Potential of Spray Drones in Agriculture* https://newswire.caes.uga.edu/story/10057/spray-drones.html. Accessed 19 July 2023.
30. J. A. Bonds, N. Pai, S. Hovinga, K. Stump, R. Haynie, S. Flack, and T. Bui, "Spray Drift, Operator Exposure, Crop Residue and Efficacy: Early Indications for Equivalency of Uncrewed Aerial Spray Systems with Conventional Application Techniques", Journal of the ASABE. Vol. 67(1) pp. 27-41, 2024.
31. P. K. Malini, R. K. Naik, V. Kumar, and Manjunatha, "Seed Planting Robotic Drone", JETIR, vol. 6, pp. 639–641, 2019.
32. E. Nonko, "Seed-Planting Drones Are Reforesting Canada With Lightning Speed", *Reasons to Be Cheerful*, 27 Feb. 2023, https://reasonstobecheerful.world/drone-tree-planting-reforesting-after-wildfires/.
33. A. Crosby, "What Is Drone Mapping?" *GeoNadir*, 19 Sept. 2022, https://geonadir.com/what-is-drone-mapping/.
34. Mary. *3D and Drone Mapping in Agriculture*. 6 July 2022, https://geopard.tech/blog/what-is-agriculture-drone-mapping/.
35. Rshaffer, "How Drones Are Growing the Agricultural Industry with Soil Analysis", *Consortiq*, 7 Oct. 2022, https://consortiq.com/uas-resources/how-drones-are-growing-the-agricultural-industry-with-soil-analysis.
36. M. T. I. Drones, "Drone for Agriculture: What It Can Do For Your Crops By: Jeremiah Burnett", *MTI Drones*, 16 Sept. 2021, https://mtidrones.com/blogs/news/drone-for-agriculture-what-it-can-do-for-your-crops.
37. M. Ayamga, B. Tekinerdogan, A. Kassahun, and G. Rambaldi, Developing a Policy Framework for Adoption and Management of Drones for Agriculture in Africa. *Technology Analysis & Strategic Management*, vol. 33, no. 8, pp. 970–987, 2021.
38. N. Alexandratos, and J. Bruinsma, *World Agriculture Towards 2030/2050: The 2012 Revision, Agricultural Development Economics Division, Food and Agriculture Organization of the United Nations* .
39. B. S. Chauhan, Grand Challenges in Weed Management, www.frontiersin.org/articles/10.3389/fagro.2019.00003
40. F. Veroustraete, "The Rise of the Drones in Agriculture", *ECronicon Agricultural Editorial*, vol. 2, no. 2, pp. 325–327, 2015.
41. D. L. K. I-Kuai Hung, D. R. Unger, and Y. Zhang, "Accuracy Assessment On Drone Measured Heights At Different Levels", p. 1.
42. S. Shuping, and J. N. Eloff, "The Use of Plants to Protect Plants and Food against Fungal Pathogens: A Review". *African Journal of Traditional, Complementary and Alternative Medicines*, vol. 14, no. 4, pp. 120–127. 2017.
43. S. Ahirwar, R. Swarnkar, S. Bhukya, and G. Namwade, "Application of Drone in Agriculture", *International Journal of Current Microbiology and Applied Sciences*, vol. 8, pp. 2500–2505, 2019.

44. Beriya, "Application of Drones in Indian Agriculture", *CSD Working Paper Series: Towards a New Indian Model of Information and Communications Technology-Led Growth and Development*, vol. 73, pp. 1–15, 2022.
45. J. Yadav, U. Chauhan, and D. Sharma, "Importance of Drone Technology in Indian Agriculture", in *Smart Village Infrastructure and Sustainable Rural Communities*, pp. 35–46, eds. Mohammad Ayoub Khan, Pushkar PraveenIGI Global, 2023.
46. J. Durfey, "Drones and Autonomous Vehicles in Agriculture", *Agricultural Technology & Production Management Program*, pp. 1–47. https://s3-us-west-2.amazonaws.com/smallgrains.wsu.edu/uploads/2019/01/James_Durphey_Drones_in_Ag.pdf
47. D. Kamthania, "Use of Drones in Agriculture", *Medicon Agriculture & Environmental Sciences*, vol. 2, pp. 49–52, 2022.
48. M. E. Karar, F. Alotaibi, A. AL Rasheed, and O. Reyad, "A Pilot Study of Smart Agricultural Irrigation using Unmanned Aerial Vehicles and IoT-Based Cloud System", *International Journal of Information Sciences Letters*, vol. 10. p. 1–11.
49. B. Azouz, H. Esmonde, B. Corcoran, and E. O'Callaghan, "Development of a Teat Sensing System for Robotic Milking by Combining Thermal Imaging and Stereovision Technique", *Computers and Electronics in Agriculture*, vol. 110, pp. 162–170, 2015.
50. M. H. Kim, J. Lee, and S. J. Lee, "Hail: Mechanisms, Monitoring, Forecasting, Damages, Financial Compensation Systems, and Prevention", *Atmosphere*, vol. 14, no. 11, p. 1642, 2023.
51. R. P. Motha, "The Impact of Extreme Weather Events on Agriculture in the United States", in *Challenges and Opportunities in Agrometeorology*, eds. Attri et al., pp. 397–407, 2011.
52. M. Constantinoiu, "Fruit-Picking Drones Can Solve the Farm Labor Shortage", *ISRAEL21c*, 17 Aug. 2022, www.israel21c.org/fruit-picking-drones-solve-the-farm-labor-shortage/.
53. "10 Major Pros & Cons of Unmanned Aerial Vehicle(UAV) Drones", https://www.avian.net.au/pros-and-cons-of-uav-surveys/ https://equinoxsdrones.com/blog/10-major-pros-cons-of-unmanned-aerial-vehicle-uav-drones. Accessed 19 July 2023.
54. "Pros and Cons for UAVs on Farms", www.farmprogress.com/technology/pros-and-cons-for-uavs-on-farms. Accessed 19 July 2023.
55. Derhab O. Cheikhrouhou, A. Allouch, A. Koubaa, B. Qureshi, M. Ferrag, and F. A. Khan, "Internet of drones security: Taxonomies, open issues, and future directions" *Vehicular Communications*, vol. 39, p. 100552, 2023.
56. "The Pros And Cons Of Drones In Agriculture – DroneGuru", 26 Nov. 2016, www.droneguru.net/the-pros-and-cons-of-drones-in-agriculture/.
57. Agrotechnomarket.com., "The Disadvantages of Agricultural Drones", *Agriculture, Technology, and Business Market*. www.agrotechnomarket.com/2017/03/the-disadvantages-of-agricultural-drones.html. Accessed 19 July 2023.
58. "Use of Drones for Agriculture: Advantages and Challenges: Daily Current Affairs", *Dhyeya IAS® – Best UPSC IAS CSE Online Coaching | Best*

UPSC Coaching | Top IAS Coaching in Delhi | Top CSE Coaching, www.dhyeyaias.com/current-affairs/daily-current-affairs/use-of-drones-for-agriculture-advantages-and-challenges. Accessed 19 July 2023.

59. "Pros And Cons Of Drones In Agriculture | Grind Drone", 25 Sept. 2017, https://grinddrone.com/info/pros-and-cons-in-agriculture.
60. High Level APET Report, "Drones On The Horizon", NEPAD, African Union, pp. 1–33, 2019.
61. B. W. Beam, and A. R. Specht, Agriculture's Eye In The Sky, Department of Agricultural Communication, Education & Leadership, Unknown, p. 1.
62. B. Lu, P. D. Dao, J. Liu, Y. He, and J. Shang, "Recent Advances of Hyperspectral Imaging Technology and Applications in Agriculture". *Remote Sensing*, vol. 12, no. 16, p. 2659, 2020.

Chapter 12

Livestock Management using the Internet of Things

Aarti Rani

12.1 INTRODUCTION

Around 7 billion people live on Earth at the moment. According to estimates between 2030 and 2050, this number will rise to between 9.4 and 10.1 billion people [1]. The rising User Demand for Animal Food (UDfAF) in this rapidly populating globe is unavoidable. On the one hand, it would become more challenging to fulfill the UDfAF due to the naturally complicated nature of cattle husbandry, the progressive decline in the world's workforce, and the rising cost of food production. Due to urban growth, the surrounding towns and villages are now integrating with the metropolis and embracing a concrete and western way of life. In order to maintain a shared society, achieve economic growth, and create a balance between lifestyle quality and smart city concept, balanced development is a fundamental prerequisite. By custom, the majority of farming communities in India's towns and villages keep domestic animals like buffalos, cows, goats, sheep, and hens (referred to as livestock) in order to generate money, primarily through the sale of milk and eggs as a home-based enterprise. Most livestock management practices are based on the classic loose housing farm concept. To maintain the quality of life of people and domestic animals, it is crucial to support farmers in their daily work and livestock management by adopting smart solutions. India is moving in the same path as other nations like South Korea, the United States, and China, who have already started to reap the benefits of IOT applications. With cutting-edge sensor monitoring and data processing technology, IOT smart solutions may serve needs by enhancing the environment, intelligent renewable energy sources, and promoting animal health. Since milk and milk products are more in-demand goods, production and milk quality must increase to remain competitive. By utilizing 'precision' agricultural techniques, IOT-based digital revolutions enable farmers to monitor, maintain, and process the crop in a 'scientific' and 'intelligent' manner [2]. Due to the size of the livestock business, it is crucial to monitor livestock and other farm animals. By combining sensors, GPS, and other technologies with a network protocol for communication,

DOI: 10.1201/9781003676102-12

a ground-breaking system for tracking livestock has been created. This monitoring equipment allows farmers to check on their farm animals from a distance. The tracker device is attached to collars and keeps track of the animals' health and also their chewing patterns, location data, and pasture management. The livestock monitoring system aids in the management of both farm animals and farm machinery. A wireless sensor network is an implementation that uses sensors on animals and transmits the data to a monitoring hub [3].

Concerns about the expansion of agricultural production to meet the needs of the world's population are becoming more widespread. Therefore, the management of livestock in agriculture is crucial for ensuring livelihood. However, farmers find it difficult to control their crops and cattle due to the growing worry over land and water resources. Aside from this, farmers continue to place a high premium on minimizing waste [4] and costs in general. The latest technological developments have a significant impact on how well and how much agricultural produce is grown. The Internet of Things (IoT) enters the picture at this point. It offers farmers the chance to improve the health of their cattle through remote monitoring and data-driven decision-making.

12.2 SMART IOT SOLUTIONS

The Internet of Things links equipment and consumer goods to the network, enabling data collection, information gathering, and smart administration of these items with software to increase efficiency. It makes it possible for new services to smartly accomplish benefits for health, safety, and the environment. A number of industries, including agriculture, are being transformed by the IoT, a rapidly developing technology. The use of IoT technology, in particular, has enormous potential benefits for livestock production. The prospective uses for the IoT in the agricultural and farming industries are reviewed briefly in this study, along with an analysis of the current problems and potential solutions. The world population is currently at over 7 billion, with an expected increase in 2030 from 9.4 billion to roughly 10 billion in 2050 [5]. The need for animal food has continued to rise, and during the past few decades, the impact of livestock farming systems on the biosphere has also increased due to the rapid development of the population. Additionally, one of the biggest obstacles to be overcome in the production of livestock is the effective management of scarce resources, such as reducing the amount of people needed, reducing the contribution of livestock to greenhouse emissions, managing water resources, and minimizing repetitious daily livestock management tasks [6]. A very promising technology that offers numerous creative options to modernize cattle husbandry is the IoT. The cattle farming industry has lately begun to pay greater attention to IoT technology because it can meet the crucial demand for compatibility across the agriculture sector with scalability and traceability [7]. It is built on the idea of identifying connected devices individually, retrieving their data, and storing it in the base station, where machine learning algorithms use it to accomplish shared objectives. Additionally, it made use of location tracking technology to track the movements of farm animals, alert authorities when they cross the farm's sensor perimeter, and assess the health and welfare of the livestock. It is believed that IoT technology can advance livestock farming methods by connecting sensor data from cattle and in particular those who are away from the farm over the internet [8]. In a situation involving cattle, an IoT network is the key component for tracking and monitoring animal behavior. An IoT network supports the transmission and receiving of livestock data while also providing accessibility to the IoT backbone and maintaining the full livestock infrastructure. As sensor and connectivity technology has improved, studies on IoT applications using drones with an emphasis on livestock management have been conducted [9]. In this field, Deep Learning (DL), Machine Learning (ML), and Artificial Intelligence (AI) tools are used to streamline data collection, processing, and real-time decision-making [10]. The IoT also provides a well-planned scheduling system for limited resources, ensuring that the IoT is used to its fullest potential and boosting productivity.

12.3 BACKGROUND STUDY

Various physiological features of animals can be measured everywhere thanks to sensing technology. Systems for intelligent cattle management employ visual object tracking, an important technique. Animal behavioral patterns can provide valuable information regarding reproduction and health [11]. All the drawbacks of conventional farming are minimized by the livestock monitoring system. Due to the farmers' ignorance of the animals' conditions, such as health problems, pregnancy reports, vaccination dates, and milk production ratios, traditional farming poses a number of concerns. They might easily avoid animal illness if they were aware of these circumstances. However farming often still uses outdated old methods. Such development in the cattle agricultural sector is not achievable without contemporary technologies. A simple thermal frame threshold that uses a topographical face as the defining boundary of an object and which performs well in overlapping circumstances is presented by the author. The separation of all animals as a result of structural changes to the topology is used as the basis for an efficient refining procedure that updates all animal placements on a regular basis. It can therefore withstand the animals' sudden movements. This camera technology uses thermal sensors to help track several objects. However, thermal camera systems require significant computational labour and can clog IoT networks [12]. The authors look into possible fixes for welfare quality assessment techniques based on sensor data in another piece of writing. A lot of improvements to PLF technology could be made. In the following article, these are discussed. The welfare quality procedure for dairy cattle is made up of 4 concepts, 12 criteria, and 31 measures. Some requirements for viable ML solutions include sleep comfort, the absence of pain brought on by therapeutic methods, and a positive human-animal bond. Social behavior expression is a field where ML systems can be improved. The time when insemination takes place, known as estrus, is one of the most important components of livestock management. Accurate estrus detection raises productivity and earnings. An intravaginal wireless device that can recognize estrus in a farm has been proposed by Andersson [13]. Detecting estrus with conventional methods is challenging and expensive. Estrus detection is made easier by wireless technology. On a larger scale, efficiency and dependability have not yet been proven. Germani et al.'s [14] proposal for continuous animal observation, processing, and analysis in farms and grasslands included a LoRa (LPWAN) based solution encompassing different environments and a web services architecture.

The networking and IoT solutions are significantly impacted by real-time field tests [15]. For livestock IoT devices, Beng et al.'s [10] experiments with Wireless Power Transfer (WPT), NFC, and WSN addressed the problem of recharging. A wireless power transfer system that fuels the animal's connected remote wireless power transfer devices with radio frequency energy. While the WSN and NFC provides internet connectivity. The WSN architecture and power transfer wattage have not been stated. The monitoring of livestock using remote sensors and Internet of Things (IoT) technologies has been the subject of numerous researches [16]. A number of IoT-based monitoring and management systems with devices linked to the cattle were surveyed by the author [17]. The main drawbacks of the systems under study are higher energy consumption, challenging interfaces, and expensive installation and maintenance, which makes farmers' adoption of these smart systems challenging. External sixteen channel wireless multi sensor PCB devices were developed by Zeng et al. [18] to collect data on the wind speed, temperature, barometric pressure, light intensity, rain, humidity, wind direction, and gas amount.

Due to the size of the livestock business, it is crucial to monitor livestock [18] and other farm animals. A breakthrough system for monitoring livestock has been designed and built using sensors, GPS, and other technologies, and it integrates all of them with a network protocol for communication. The farmers may remotely check on their farm animals thanks to this monitoring device. The tracker gadget is fastened to collars and monitors the wellbeing of the animals as well as their chewing habits, geographic information, and pasture management.

12.4 IOT ENABLED LIVESTOCK MONITORING MANAGEMENT

The platform for IoT-enabled livestock [19] monitoring may prove invaluable for the farming of farm animals. IoT-enabled livestock control solutions provide records on a range of animal health variables. Animal position, temperature, blood pressure, and heart rate can all be tracked with a wearable collar or tag equipped with sensors. These wearable IoT devices wirelessly transmit each livestock's records from the farm to an important component known as the IoT gateway.

Small and light in weight, the tracking gadget is designed to not feel bulky on an animal's body. The equipment has sensors and a GPS tracker so that the movement and fitness of the animals can be tracked around-the-clock. The device's behavioral monitoring feature can help farmers determine what their cattle are eating. In relation to the hardware, there are cellular applications, web applications, or IoT dashboards. Farmers can stabilize the areas where the cattle can pass freely by creating geofencing-enabled digital barriers with the aid of the device-related software. If the cattle travel through the hurdles indicated, the tracker transmits signals to the mobile programme. Additionally, it transmits signals if the cattle consume or absorb something harmful or improper. Farmers are able to protect their animals from health problems in this way. The device may operate over LPWAN, which is ubiquitous, ensuring proper connectivity. Depending on the application, different LPWAN protocols are chosen. The growth and awareness of current technology have significantly improved farm animal or livestock health. A few benefits include high earnings and organized farm animal management.

12.5 IOT-ENABLED LIVESTOCK MANAGEMENT BENEFITS

Sectors like agriculture and livestock are undergoing significant change as a result of IoT solutions and smart devices. It is really difficult for us to envisage a world without the Internet of Things when everything is connected to one another around us. The internet of things has led to changes in the livestock industry. Farmers can now monitor their cattle without having to haul around bulky devices by using battery-powered sensors and smart technology. By using IoT-based livestock monitoring, ranchers may track their cattle in real-time. It helps with a variety of problems the cattle industry has. Direct engagement between the employees and each of the animals on farms nowadays is impracticable due to the huge number of animals present. In addition to providing trustworthy data for each animal on the farm, IoT-enabled surveillance for livestock makes it simpler to keep an eye on all of the animals from a distance. Some of the benefits are shown in Figure 12.1.

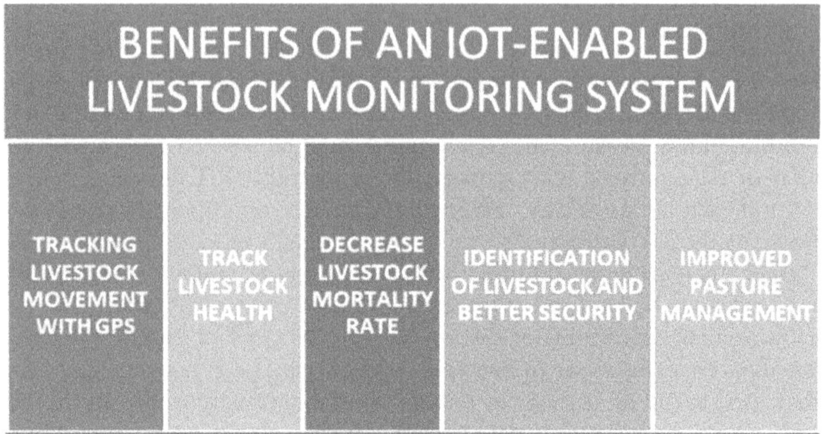

Figure 12.1 Benefits of IoT in Livestock Monitoring System.

- Farmers can keep an eye on dietary changes and feeding habits.
- Farmers can cure animals right away thanks to real-time monitoring of their health and performance, which also helps to stop the spread of disease.
- Farmers can track grazing creatures.
- Farmers can compile and research historical data to discover the traits that make farm animals fit.
- Farmers can improve breeding methods.
- Veterinarians, nutritionists, and other professionals can access real-time data as well as historical records via cloud-based integration and dashboard solutions.

12.6 INTERNET OF THINGS IN ANIMAL HUSBANDRY

An extensive network of sensor nodes makes up the IoT for animal husbandry, which is a network for monitoring. It collects information using a range of sensors to help farmers spot problems early and precisely pinpoint the specific poultry and livestock that require treatment [20, 21]. The use of RFID technology, a non-contact rapid identification device, in modern aquaculture is growing. The management system for animal husbandry and the animal husbandry Internet of Things aims to realize the fusion of the internet and electronic ear tag technology of unique identification of animal identity on the basis of compiling priceless experience from various animal husbandry and animal husbandry processes [22]. It follows the national agricultural development policy of 'integrating resources and system integration.' The key is to actually implement agricultural husbandry and animal husbandry around the needs of standardization, automation, and the development of high-quality safety systems. While using an IoT management system for animal husbandry, farmers can get data from readers regarding the identity of their livestock, including breed, date, and location of birth; whether to avoid outbreaks; dietary information, including quantity, kind, and frequency of feeding; and data about epidemic prevention [23]. By supporting livestock management organizations at all levels, in swiftly and easily comprehending the production data of meat, eggs, milk, and various other animal products under their jurisdiction, the poultry and livestock breeding supervision system may support government decision-making [24]. A centralized farming strategy with substantial managerial complexity and densities is used in modern poultry and livestock farming. Modern Internet of Things technology was incorporated into the manufacturing management procedure while also considering the characteristics of the development of the livestock and poultry industries in order to address these difficulties. Automation, standardization, and intellectualization in animal husbandry and aquaculture have raised the bar for production management even further [25]. The Internet of Things platform can examine the data remotely; conduct intelligent analysis, and implement data notification and automatic control features since the Internet of Things acquiring tool transfers the data there in real time [26].

12.7 IMPACT OF THE IOT ON AGRICULTURE

The phrase 'smart farming' is a fantastic way to define agriculture in this Fourth Industrial Revolution. Technologies in agriculture are currently quite exciting due to the merging of the Fourth Industrial Revolution's innovations with what is also referred to as the 'Second Green Revolution.' Producers, cultivation, and agribusiness are facing unimaginable disruption on a global scale. Agricultural IoT, also known as smart farming, is a sometimes underappreciated use case for the IoT since it doesn't fall under any of the well-known subcategories like health, mobility, or industrial IoT.

- Scientists utilize the IoT to investigate how genes and microclimates impact crop production to enhance crop quality and productivity [27].
- By deploying smart, connected IoT projects, farmers may exploit the huge volumes of data collected on their farms.
- One option to increase production and guarantee food quality across the supply chain is to follow farming activities from harvest to delivery using IoT track-and-trace technology and satellite photos.
- Future improvements in product superiority, crop yield, resource management, and expenditure management, to mention a few, are

just some of the ways that the agricultural IoT has the potential to transform food production and farming.

- The agricultural market will see an average 20% yearly growth in IoT device installations, predicts BI Intelligence. By producing 7,340 kilograms of cereal per hectare (2.5 acres) of land, on average, as opposed to the global average of 3,851 kilograms per hectare, American farmers have already shown the success of smart farming approaches.

12.8 SENSORS IN AGRICULTURE

A sensor is a piece of technology that recognizes and detects specific inputs like brightness, movement, pressure, heat, or moisture and responds by translating those inputs into representations or signals that people can comprehend. Different types of sensors are utilized in agriculture, which makes the adoption of smart agriculture necessary.

Every aspect of conventional farming practices can be drastically altered by integrating the most recent sensing and IoT technologies [28]. The IoT and wireless sensors may now be seamlessly integrated into smart agriculture to take it to previously inconceivable heights. Many conventional farming problems, like responding to droughts, optimizing yields, finding suitable land, irrigation [29,30], and pest control, can be improved with the use of IoT. Sensors are mostly employed in agriculture for quality assurance, safety, and traceability, for plant and animal perception, and environmental information monitoring. At present, the three kinds of sensors most frequently employed in agriculture are chemical sensors, biosensors, and micro-electro-mechanical sensors. Physical property sensors convert signals through sensor-sensitive physical changes, whereas biosensors provide information based on the reaction to the outside world. Recent technical advancements like micro-electro-mechanical sensors have exceptional properties like low power utilization and high reliability [31]. Figure 12.2 shows some kinds of sensors that are used.

12.9 CONCLUSION

Urban cities are growing and absorbing neighboring towns and villages as their populations rise. Therefore, significant collaboration between urban and rural communities is essential for the future of smart cities. The IoT phenomena in smart cities are important for boosting the national ecology since it may simultaneously enhance quantity and quality. The study offered many IOT-based tools for farmers to efficiently organize their daily care of cattle and manage their animals. IoT hardware and software are now present in practically all spheres of society. The device swiftly and correctly detects animal health and physiological status as well as cattle activity. One of the prospective infrastructures, known as the IOT, which

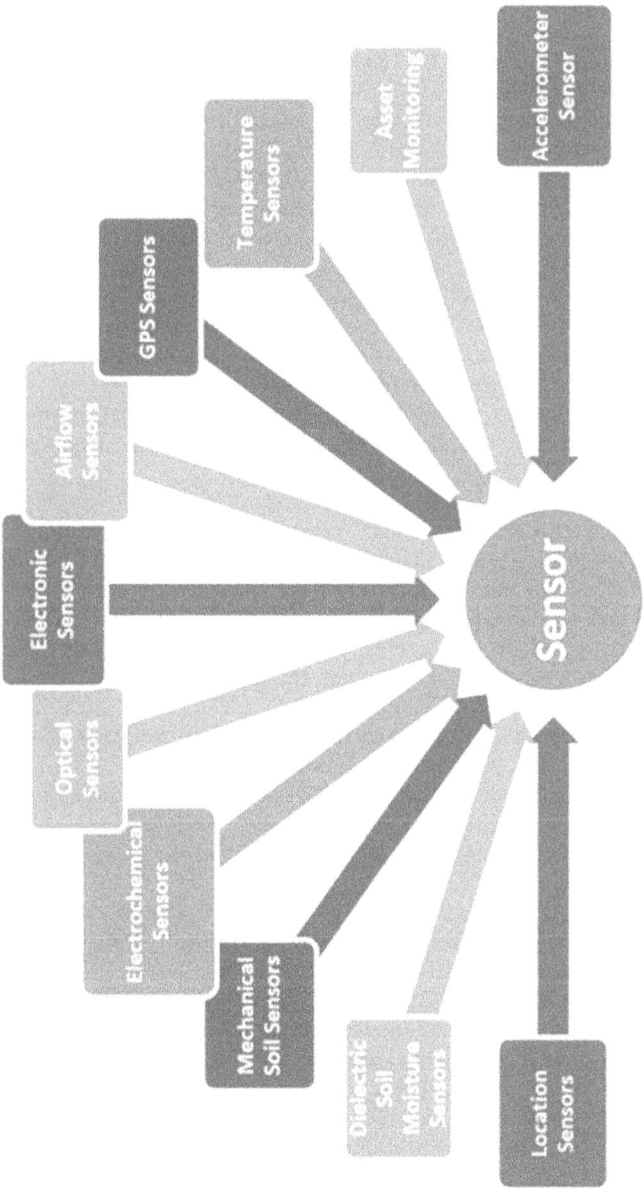

Figure 12.2 Types of Sensors.

has limitless potential in terms of hardware, software, communication, and cloud systems, advances this activity.

REFERENCES

1. J. Uillen, F. Natale, N. Carvalho, J. Casey, J. Hofherr, J.-N. Druon, G. Fiore, M. Gibin, A. Zanzi, and J. T. Martinsohn, "Global seafood consumption footprint," *Ambio*, vol. 48, pp. 111–122, 2019.
2. E. Navarro, N. Costa, and A. Pereira, "A systematic review of IoT solutions for smart farming," *Sensors*, vol. 20, p. 4231, 2020.
3. W. Sweden, Five IoT Smart Agriculture Use Cases www.iotforall.com/five-iot-smart-agriculture-use-cases accessed on March 12, 2024.
4. Greenhouse smart application project that reduces waste and improves workforce efficiency at farms in Spain, "Advantech Co-Creating the Future of the IoT World," n.d. www.advantech.com/en/resources/case-study/greenhouse-smart-application-project-that-reduces-waste-and-improves-workforce-efficiency-at-farms-in-spain
5. United Nations, Department of Economic and Social Affairs, Population Division. *World Population Prospects 2019: Highlights*, United Nations, 2019.
6. K. Dineva, and T. Atanasova, "Design of scalable IoT architecture based on AWS for smart livestock," *Animals*, vol. 11, p. 2697, 2021.
7. M. A. Alanezi, M. S. Shahriar, M. B. Hasan, S. Ahmed, Y. A. Sha'aban and H. R. E. H. Bouchekara, "Livestock management with unmanned aerial vehicles: A review," *IEEE Access*, vol. 10, pp. 45001–45028, 2022.
8. L. F. Si, M. Y. Li, L. He, "Farmland monitoring and livestock management based on internet of things," *Internet of Things*, vol. 19, p. 100581, 2022.
9. M. Gharibi, R. Boutaba and S. L. Waslander, "Internet of Drones", *IEEE Access*, vol. 4, pp. 1148–1162, 2016.
10. M. S. Mahmud, A. Zahid, A. K. Das, M. Muzammil and M. U. Khan, "A systematic literature review on deep learning applications for precision cattle farming", *Computers and Electronics in Agriculture*, vol. 187, Aug. 2021.
11. W. Kim, Y. B. Cho, and S. Lee, "Thermal sensor-based multiple object tracking for intelligent livestock breeding," *IEEE Access*, vol. 5, pp. 27453–27463, 2017.
12. F. M. Molina, C. C. P. Mar´ın, L. M. Moreno, E. I. A. Buend´ıa, and D. C. P. Mar´ın, "Welfare quality® for dairy cows: Towards a sensor based assessment," *Journal of Dairy Research*, vol. 87, no. S1, pp. 28– 33, 2020.
13. L. M. Andersson, H. Okada, R. Miura, Y. Zhang, K. Yoshioka, H. Aso, and T. Itoh, "Wearable wireless estrus detection sensor for cows," *Computers and Electronics in Agriculture*, vol. 127, pp. 101–108, 2016.
14. L. Germani, V. Mecarelli, G. Baruffa, L. Rugini, and F. Frescura, "An IoT architecture for continuous livestock monitoring using loralpwan," *Electronics*, vol. 8, no. 12, p. 1435, 2019.

15. S. S. AbouEmira, K. Y. Youssef, and M. Abouelatta, "Power efficient routing protocol (aodv) for iot based livestock applications," in *2019 14th International Conference on Computer Engineering and Systems (ICCES)*, pp. 407–412, IEEE, 2019.
16. L. T. Beng, P. B. Kiat, L. N. Meng, and P. N. Cheng, "Field testing of iot devices for livestock monitoring using wireless sensor network, near field communication and wireless power transfer," in *2016 IEEE Conference on Technologies for Sustainability (SusTech)*, pp. 169–173, IEEE, 2016.
17. T. Vigneswari et al., "Smart IoT cloud based livestock monitoring system: A survey," *Turkish Journal of Computer and Mathematics Education (TURCOMAT)*, vol. 12, no. 10, pp. 3308–3315, 2021.
18. J. Zhang, F. Kong, Z. Zhai, S. Han, J. Wu, and M. Zhu, "Design and development of iot monitoring equipment for open livestock environment," *International Journal of Simulation: Systems, Science and Technology*, vol. 17, no. 26, pp. 2–7, 2016.
19. Livestock monitoring market size, trends and global forecast to 2032, "Global Market Research Reports & Consulting | The Business Research Company," n.d. www.thebusinessresearchcompany.com/report/livestock-monitoring-global-market-report
20. GoDaddy security, "GoDaddy Security – Access Denied," n.d. www.analyticsinsight.net/top-5-agriculture-drones-start-ups-to-know-in-2021/
21. N.-A. M. Al-Awkally, H. K. Ibrahim, and A. Samad, "Antipsychotic combinations for psychiatric disorders," *BULLET: Jurnal Multidisiplin Ilmu* vol. 1, no. 1, pp. 49–50, 2022.
22. M. Al-Bahrani, and A. Cree, "In situ detection of oil leakage by new self-sensing nanocomposite sensor containing MWCNTs," *Applied Nanoscience*, vol. 11, no. 9, pp. 2433–2445, 2021.
23. S. H. Ismael, S. W. Kareem, and F. H. Almukhtar, "Medical image classification using different machine learning algorithms," *AL-Rafidain Journal of Computer Sciences and Mathematics*, vol. 14, no. 1, pp. 135–147, 2020.
24. A. Samad, M. Hamza, A. Muazzam, A. Ahmer, S. Tariq, M. J. Shahid, and F. U. Din, "Overview of bacterial diseases in poultry and policies to control disease and antibiotic resistance," *BULLET: Jurnal Multidisiplin Ilmu*, vol. 1, no. 1), pp. 19–25, 2022.
25. A. Samad, M. Hamza, A. Muazzam, and M. K. Harahap, "Role of artificial intelligence in livestock and poultry farming," *Sinkron: jurnaldanpenelitianteknikinformatika*, vol. 7, no. 4, pp. 2425–2429, 2022.
26. B. N. Mohammed, F. H. Al-Mukhtar, R. Z. Yousif, and Y. S. Almashhadani, "Automatic classification of Covid-19 chest X-ray images using local binary pattern and binary particle swarm optimization for feature selection," *Cihan University-Erbil Scientific Journal*, vol. 5, no. 2, pp. 46–51, 2021.
27. IoT based weather monitoring system for micro-climate forecasting, "Benchmark Labs," 2022, Apr. 6. www.benchmarklabs.com/blog/micro-climate-forecasting/
28. R. K. Kodali, V. Jain, and S. Karagwal, "IoT based smart greenhouse," In *2016 IEEE Region 10 Humanitarian Technology Conference (R10-HTC)*, pp. 1–6, IEEE, Dec. 2016.

29. "What is smart irrigation?," HydroPoint, n.d.. www.hydropoint.com/what-is-smart-irrigation/
30. What Is Smart Irrigation – Learn About Smart Watering Technology, 2022, www.gardeningknowhow.com/garden-how-to/watering/what-is-smart-irrigation.htm
31. J. Li, M. R. Guo, and L. L. Gao, "Application and innovation strategy of agricultural Internet of Things," *Transactions of the Chinese Society of Agricultural Engineering*, vol. 31, no. 2, pp. 200–209, 2015.

Chapter 13

Agricultural Supply Chain

Gaurab Kumar Sharma and Rajesh Kumar Ghai

13.1 INTRODUCTION

The production, processing and distribution of food, fiber and other products from plant and animal husbandry are collectively referred to as the agricultural sector. Because it provides products needed for human consumption and subsistence, it is an important sector of any economy. The agricultural sector is important on many levels. Firstly, it satisfies the primary nutritional needs of the population. It also creates employment opportunities, especially in rural areas, and supports the country's overall economic development. Agribusiness is an important supplier of raw materials for many industries, including textiles, pharmaceuticals and bio-fuels. In addition, the agricultural sector contributes significantly to sustainable development by supporting environmental protection and resource management (Swamy & Dharani, 2016).

Agriculture remains a high priority sector in the Indian economy, supporting approximately 58% of the country's population. Analysis of food supply and demand shows that despite a 50% increase in crop production over the past decade, the rate of increase in food production will not be sufficient to feed the projected growing world population of around 10 billion by 2050 (World Bank, 2019). Therefore, the productivity of the agricultural industry needs to be accelerated (Forbes, 2021).

Technology has become critical to powering all of the world's major economies. All sectors, including agriculture, are innovating to tackle complex problems. India, one of the largest economies in the world, is no exception and has seen major technological transformation over the past two decades (IndiaAI, 2021).

13.2 INDIAN AGRICULTURE SYSTEM

The agricultural sector in India contributes significantly to the country's economy and provides livelihood for around 58% of the population. With a share of 11.9% in global agricultural gross value added, the country also

DOI: 10.1201/9781003676102-13

has a prominent position in the global economy, ranking second after China. Despite its importance, the sector faces several challenges along the value chain. These challenges include a lack of technology integration, which can lead to unsustainable practices, food waste, and environmental costs. The challenges are summarized in the Figure 13.1.

The challenges mentioned above clearly indicate the need for disruptive interferences which allow for interventions via technological solutions. The agricultural system needs to adapt to a holistic approach which can be built upon indigenous and traditional farming knowledge by integrating it with transformative smart farming practices.

1	Small and marginal farmers (86% of farmers own <2 ha of land, causing unsustainable farm incomes and poverty)
2	Unsustainable farming practices, resulting in soil degradation and water stress
3	Unorganised and fragmented systems with existence of multiple levels of intermediaries and middlemen across the agricultural value chain
4	Limited access to technology, inputs, credit, capital and market, etc.
5	High-input, resource-intensive farming systems causing massive deforestation, water scarcities, soil depletion and high levels of greenhouse gas emissions
6	Lack of datasets at farm, farmer and sector levels, leading to higher cost of services
7	Gaps in market linkages, challenges in price discovery for farmers and price volatility in the market
8	Challenges in financial and digital inclusivity
9	Lack of food processing, logistics and warehousing infrastructure close to farm gates, increasing wastage
10	Poor farm mechanisation due to affordability challenges

Figure 13.1 Source: Market survey and PwC analysis.

13.2.1 Emerging Need for AI in Agriculture

In recent years, the agricultural industry has experienced a technological revolution in which the integration of AI has become increasingly common. This is due to the need to improve agricultural efficiency, increase yields, and reduce waste. AI has the potential to revolutionize the industry by giving farmers the tools to make data-driven decisions, improve resource management and optimize crop yields. One of the main reasons for the need for AI in agriculture is the growing world population and the need to produce more food to feed it. AI can help increase crop yields, reduce waste and optimize resource use, allowing farmers to produce more food with fewer resources (Ganesh Kumar et al., 2023).

Another factor driving the need for AI in agriculture is the increasing complexity of modern farm operations. Today's farmers face numerous challenges, including climate variability, pests and diseases, and changing market conditions. AI can help farmers address these challenges by providing information about crop health, soil moisture levels and weather conditions, predicting crop yields and identifying potential problems before they occur (Tu, 2018).

The adoption of AI technologies will pave the way for higher production with optimal use of available resources, facilitating predictive analysis, crop health management, improving quality and traceability, and so forth. The adoption of innovative and transformative smart farming practices in the country is gradually emerging as a major trend (India Global Business, 2022).

From the global market point of view, smart agricultural systems and technologies, including AI and Machine Learning (ML), are showing remarkable growth with the investment and expenditure trends expected to triple by 2025 to USD 15.3 billion, out of which AI technologies alone are projected to grow at a Compound Annual Growth Rate (CAGR) of 25.5%. Within AI interventions, Internet of Things (IoT) enabled Agricultural (IoTAg) monitoring is considered to be the fastest-growing technology segment, and projected to be worth USD 4.5 billion by 2025 (Ganesh Kumar et al., 2017).

The global market size for AI in agriculture stood at USD 852.2 million in 2019 and is expected to reach USD 8,379.5 million by 2030, exhibiting a CAGR of 24.8% during the forecast period (2020–2030). This market growth is propelled by increasing penetration of IoT in the agriculture industry with implementation of data generation through sensors and aerial images for crops, leading to an increase in crop productivity through deep-learning technology (Ghahremani-Nahr et al., 2021).

Major global trends necessitating the need for increasing farm productivity are shown in Figure 13.2.

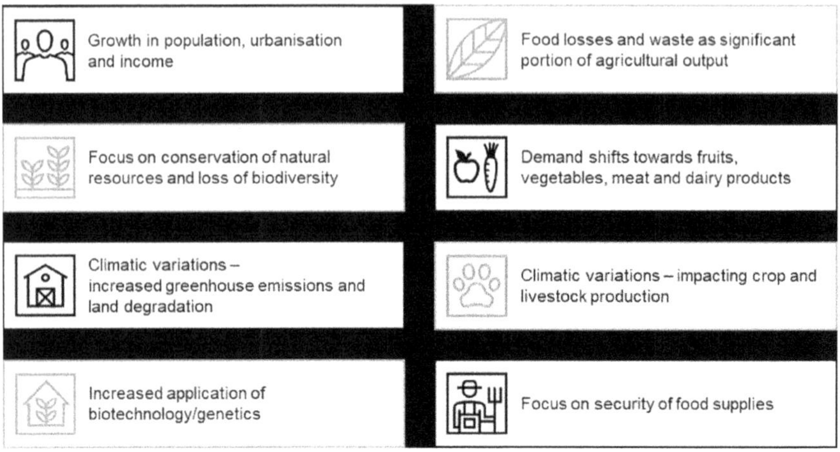

Figure 13.2 Source: UN DESA, Fitch analysis and PwC analysis.

Note: 68% of the global population is projected to live in urban areas by 2050 as the cumulative global GDP witnesses a 130% growth between 2016–2050. Around 14% of the food produced in the world is lost between farm gate and retail points, and by 2025, 35.3% of household budgets will be spent on food and drinks, up from 33.2% in 2005.

13.2.2 Agricultural Supply Chain

The supply chain is a crucial part of the agricultural industry, as it is responsible for ensuring that agricultural products are transported from the field to the end consumer in a timely, efficient, and profitable manner. A well-functioning supply chain is essential to ensuring farmers get their produce to market and receive a fair price for their crops or livestock (Chen et al., 2013a).

One of the key functions of the agricultural supply chain is to ensure that produce is transported from the farm to processing plants, where it can be turned into value-added products such as dairy, meat or processed foods. From there, products are shipped to distribution centers and then to retailers or directly to consumers. Any disruption or inefficiency in this process can lead to increased costs and reduced profitability for farmers (Alyahya et al., 2016).

In addition to the economic importance of the supply chain, it also plays a crucial role in ensuring the safety and quality of food. Proper handling, storage and transportation of produce are vital to ensure they remain fresh and safe for consumption. Any error or failure in the supply chain can lead to spoilage or contamination, which can have serious implications for consumer health and safety (Ganesh Kumar et al., 2017). Given the importance of the agricultural supply chain, there is increasing focus on improving its efficiency, sustainability and resilience. The integration of AI and IoT technologies is playing an increasingly important role in this, enabling farmers and other stakeholders to collect and analyze data on everything from crop yields and weather conditions to routes, transport and logistics (Tu, 2018).

13.2.2.1 Introduction of AI and IoT Technologies to the Agricultural Supply Chain

The agricultural supply chain is undergoing a major transformation with the introduction of AI and IoT technologies. This integration enables better data collection, optimized processes for operations such as crop production or inventory management, greater transparency in pricing decisions, and more efficient decision-making within supply chains (Abu et al., 2022). The automation opportunities offered by these new technologies can result in cost savings through increased efficiencies throughout the system, while reducing waste products at various stages of the system. In addition, this coupling allows for a higher level of customization, which benefits both manufacturers and consumers. The integration of AI and IoT technologies is transforming the agricultural supply chain by providing more accurate real-time data and enabling more efficient and effective decision-making. IoT devices such as sensors and drones can collect data on the weather, soil moisture, crop growth and animal health, amongst other things (Ghahremani-Nahr et al., 2021). AI algorithms can then analyze this data and provide insights into crop yields, disease outbreaks and supply chain optimization (Tu, 2018).

An example of the use of AI and IoT technologies in agriculture is precision farming. Precision farming uses data analytics and IoT sensors to optimize crop yields and reduce waste. Farmers can use IoT sensors to monitor soil moisture, temperature, and nutrient levels, and use AI algorithms to determine optimal planting and harvesting times (Patel et al., 2021).

13.2.2.2 Application of AI and IoT in Agricultural Logistics and Transportation

Another area where AI and IoT can be beneficial in the agricultural supply chain is in logistics and transportation. By integrating IoT sensors and AI algorithms into the logistics process, it is possible to optimize the route planning and delivery schedules, reducing the time and costs associated with transporting goods (Alyahya et al., 2016). The application of AI and IoT in agricultural logistics and transportation has the potential to significantly improve the efficiency of the supply chain, reduce waste, and increase profits for farmers and other stakeholders. Here are some ways AI and the IoT can be used in this context:

1 Route optimization: AI algorithms can be used to optimize logistics and transportation routes, reducing transportation time and costs. By analyzing real-time traffic and weather data, AI systems can suggest the most efficient route for transportation, helping farmers to save both time and money (Kamble et al., 2020).
2 Inventory management: IoT sensors can be used to monitor inventory levels in real-time, providing accurate and up-to-date information on the availability of products. This can help farmers to better manage their inventory, avoiding shortages and overstocking (Chen et al., 2013a).
3 Predictive maintenance: IoT sensors can also be used to monitor the condition of transportation equipment such as tractors and trucks, providing real-time data on their performance. This data can be used to predict when maintenance is needed, reducing downtime and preventing breakdowns that can lead to delays and loss of product (Ganesh Kumar et al., 2023).
4 Supply chain transparency: AI and the IoT can be used to provide greater visibility and transparency across the supply chain. For example, block-chain technology can be used to track the movement of products from farm to table, providing consumers with information on the origin of the product and how it was produced (Barratt & Oke, 2007).

AI and the IoT can also be used to track inventory levels and monitor the quality of agricultural products at every stage of the supply chain. This helps to reduce food waste by ensuring that perishable goods are transported and stored under optimal conditions (Ganesh Kumar et al., 2017).

13.3 INTERNET OF THINGS APPLICATIONS FOR SMART FARMING

A Review (2021) by S. Patel et al. reviews the use of IoT technologies in agriculture and provides examples of IoT applications for precision agriculture, livestock monitoring, and environmental monitoring. It also discusses the challenges associated with implementing these technologies, such as data security and the need for standardized protocols.

13.3.1 Livestock Monitoring

IoT sensors can be used to monitor the health and behavior of individual animals, allowing farmers to identify potential health issues and provide timely interventions. IoT sensors can be utilized to monitor the health and behavior of individual animals in the agricultural supply chain. These sensors can track a variety of parameters such as body temperature, heart rate, and animal behavior patterns, such as feeding and drinking habits, which can provide insight into potential health issues or stressors. This data can help farmers adjust their management practices, such as altering feed or watering schedules, to improve the overall health and welfare of the animals (Abu et al., 2022).

13.3.2 Environmental Monitoring in Agriculture

This is crucial, particularly in relation to weather and climate conditions, as it provides farmers with valuable insights and data that can inform their decision-making processes. Weather and climate conditions have a significant impact on crop growth and yield, and being able to monitor these conditions allows farmers to optimize their practices and increase productivity (Swamy & Dharani, 2016).

One important aspect of environmental monitoring is **weather forecasting.** By collecting data on temperature, humidity, precipitation, wind speed, and other weather variables, farmers can make accurate weather forecasts and be prepared for extreme weather events, such as hurricanes, droughts, or floods, by implementing mitigation strategies and reducing crop losses (Tu, 2018).

13.3.3 Climate Monitoring

This is also critical in agriculture, as it provides insight into long-term trends and changes that can affect crop growth and productivity. By monitoring climate variables, such as temperature, rainfall patterns, and soil moisture levels, farmers can identify areas where crops may be vulnerable to drought or other climate-related stressors (Forbes, 2021). Environmental monitoring in agriculture can also include the monitoring of soil health and water quality. By analyzing soil and water samples, farmers can determine the nutrient levels in their soil, identify any potential contaminants or pollutants, and take appropriate measures to maintain soil health and water quality (India Global Business, 2022).

13.4 CHALLENGES AND OPPORTUNITIES

The integration of AI and IoT technologies in the agricultural supply chain offers significant opportunities for improving efficiency and productivity, reducing waste, and enhancing sustainability. However, there are also a number of challenges that need to be addressed in order to fully realize the potential of these technologies.

One of the key challenges is data privacy and security. With the increasing use of IoT sensors and devices, there is a risk that sensitive data could be compromised or misused. Farmers and other stakeholders need to be

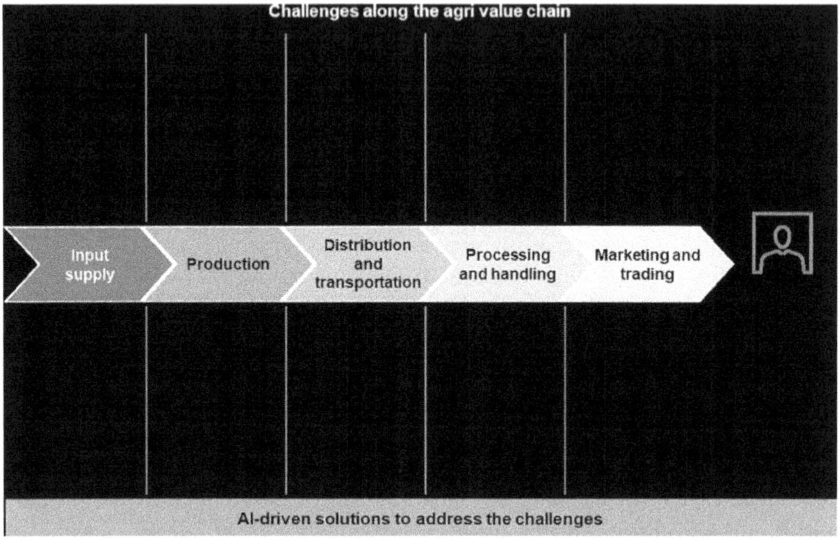

Figure 13.3 Source: Inc42 and PwC analysis.

confident that their data is being collected and stored securely, and that appropriate measures are in place to protect their privacy (Swamy & Dharani, 2016; Patel et al., 2021).

Another challenge is cyber security. As IoT devices become more prevalent in agriculture, there is a risk that they could be targeted by hackers and other malicious actors. It is essential that farmers and other stakeholders are aware of the risks and take steps to secure their systems and devices (Tu, 2018; Ghahremani-Nahr et al., 2021).

The integration of AI and IoT technologies also requires specialized skills and expertise. Farmers and other stakeholders need to be able to develop and maintain these systems, as well as analyze and interpret the data they generate. This requires a combination of technical skills, such as programming and data analysis, as well as an understanding of agricultural practices and processes (Figure 13.3) (Alyahya et al., 2016; Kamble et al., 2020).

13.4.1 Understanding the Bottlenecks

Some of the bottlenecks that hinder the adoption and implementation of AI in the Indian agriculture sector are:

1 Limited access to data: Data collection and management in the Indian agricultural system is still fragmented, unstructured and often incomplete. This poses a significant challenge for the development of

AI algorithms, which require large and diverse data sets to operate effectively (Abu et al., 2022; Swamy & Dharani, 2016).

2 Shortage of skilled workers: The development and implementation of AI solutions requires a skilled workforce with expertise in areas such as machine learning, data analysis and computer programming. However, there is a shortage of such professionals in the Indian agricultural sector, which makes it difficult to build and maintain AI-supported systems (Ganesh Kumar et al., 2023; Patel et al., 2021).
3 Infrastructure challenges: Most farmers in India work on small farms with limited access to basic infrastructure such as electricity, internet connectivity and cellular networks. This lack of infrastructure creates barriers to deploying AI-powered tools and technologies that require stable and reliable network connectivity (Tu, 2018; Kamble et al., 2020).
4 Implementation costs: The introduction of AI in the agricultural sector requires significant investments in technology, hardware and software. Small farmers, who form the backbone of India's agricultural system, may not have the financial means to invest in such systems, making it difficult to access and benefit from AI-enabled solutions (Ghahremani-Nahr et al., 2021; Forbes, 2021).
5 Regulatory challenges: The use of AI in the agricultural sector is subject to regulations and policies that need to be developed and implemented to ensure their safe and ethical use. However, the lack of a clear regulatory framework for AI in agriculture poses a challenge for its uptake and integration into the sector (India Global Business, 2022).
6 Lack of knowledge: Farmers in India, especially in rural areas, may be unaware of the potential benefits of AI and may have misconceptions or concerns about the technology. This lack of awareness and education about the benefits of AI in agriculture hampers its uptake and implementation (Patel et al., 2021; Alyahya et al., 2016).

13.5 CONCLUSION

The future of AI and IoT in agriculture is bright, with these technologies poised to revolutionize the industry and help to address some of the key challenges facing farmers and other stakeholders. One of the key areas where AI and IoT technologies are likely to have a significant impact is in precision agriculture. By using sensors and other IoT devices to collect data on soil moisture, temperature, and nutrient levels, farmers will be able to optimize their use of resources such as water and fertilizer, reducing costs and enhancing sustainability (Ghahremani-Nahr et al., 2021; Abu et al., 2022). In addition, AI algorithms will enable farmers to analyze this data and make more informed decisions about crop management, such as when to plant and harvest crops, and how to prevent or mitigate pest and disease outbreaks. AI and IoT technologies will also have a major impact on livestock management. By using sensors to monitor the health and behavior of individual animals, farmers will be able to identify potential health issues and provide timely interventions, improving animal welfare and reducing the risk of disease outbreaks (Ganesh Kumar et al., 2023; Kamble et al., 2020).

Another area where AI and IoT technologies will have a significant impact is in supply chain management. By using IoT devices to track and monitor the movement of goods, farmers will be able to improve the efficiency of supply chain management, reducing waste and improving profitability. In addition, AI algorithms will enable farmers to analyze this data and make more informed decisions about supply chain management (Tu, 2018; India Global Business, 2022).

However, there are also a number of challenges that need to be addressed in order to fully realize the potential of AI and IoT technologies in agriculture. One of the key challenges is the need for standardized protocols to ensure interoperability between different devices and systems. Another challenge is the need to develop specialized skills and expertise in order to develop and maintain these systems, as well as analyze and interpret the data they generate. In addition, there are concerns around data privacy and cyber security, which need to be addressed in order to ensure that sensitive data is collected and stored securely (Patel et al., 2021; Forbes, 2021). Despite these challenges, the future of AI and IoT in agriculture is bright. By using these technologies to optimize resource use, improve livestock management, and enhance supply chain management, farmers will be able to reduce costs, increase productivity, and enhance sustainability. As these technologies continue to evolve and become more widespread, it is likely that they will play an increasingly important role in shaping the future of the agricultural industry.

REFERENCES

Abu, N. S., Bukhari, W. M., Ong, C. H., Kassim, A. M., Izzuddin, T. A., Sukhaimie, M. N., Norasikin, M. A., & Rasid, A. F. A. (2022). Internet of Things applications in precision agriculture: A review. *Journal of Robotics and Control (JRC)*, *3*(3), 338–347. https://doi.org/10.18196/jrc.v3i3.14159

Alyahya, S., Wang, Q., & Bennett, N. (2016). Application and integration of an RFID-enabled warehousing management system–A feasibility study. *Journal of Industrial Information Integration*, *4*, 15–25. https://doi.org/10.1016/j.jii.2016.08.001

Barratt, M., & Oke, A. (2007). Antecedents of supply chain visibility in retail supply chains: A resource-based theory perspective. *Journal of Operations Management*, *25*(6), 1217–1233. https://doi.org/10.1016/j.jom.2007.01.003

Chen, J. C., Cheng, C. H., & Huang, P. B. (2013a). Supply chain management with lean production and RFID application: A case study. *Expert Systems with Applications*, *40*(9), 3389–3397. https://doi.org/10.1016/j.eswa.2012.12.047

Forbes. (2021). 10 Ways AI Has the Potential to Improve Agriculture in 2021. www.forbes.com/sites/louiscolumbus/2021/02/17/10-ways-ai-has-the-potential-to-improve-agriculture-in-2021/?sh=5434484e7f3b

India Global Business. (2022). Digitalisation has brought a fresh breath to India's agricultural culture. www.indiaglobalbusiness.com/analyses/in-focus/digitalisation-has-brought-a-fresh-breath-to-indias-agricultural-culture.

IndiaAI (2021) https://indiaai.gov.in/article/ai-impact-on-india-how-ai-will-transform-indian-agriculture

Ganesh Kumar, C., Jena, S. K., Sivakumar, A., & Nambirajan, T. (2023). Artificial intelligence in agricultural value chain: Review and future directions. *Journal of Agribusiness in Developing and Emerging Economies*, *13*(3), 379–398. https://doi.org/10.1108/jadee-07-2020-0140

Ganesh Kumar, C., Pachayappan, M., &Madanmohan, G. (2017). Agri-food supply chain management: Literature review. *Intelligent Information Management*, *9*(2), 68–96. https://doi.org/10.4236/iim.2017.92004, Available at SSRN: https://ssrn.com/abstract=3093241 or http://dx.doi.org/10.2139/ssrn.3093241

Ghahremani-Nahr, J., Nozari, H., &Sadeghi, M. E. (2021). Green supply chain based on artificial intelligence of things (AIoT). *International Journal of Innovation in Management Economics and Social Sciences*, *1*(2), 56–63. https://doi.org/10.52547/ijimes.1.2.56

Kamble, S. S., Gunasekaran, A., & Sharma, R. (2020). Modeling the blockchain enabled traceability in agriculture supply chain. *International Journal of Information Management*, *52*, 101967. https://doi.org/10.1016/j.ijinfomgt.2019.05.023

Patel, S., Kumar, N., & Gohil, A. M. (2021). Internet of Things applications for smart farming: A review. *Advances in Agriculture*, *2021*, Article ID 4567890. https://doi.org/10.1155/2021/4567890

Swamy, V., & Dharani, M. (2016). Analyzing the agricultural value chain financing: approaches and tools in India. *Agricultural Finance Review*, *76*(2), 211–232. https://doi.org/10.1108/afr-11-2015-0051

Tu, M. (2018). An exploratory study of Internet of Things (IoT) adoption intention in logistics and supply chain management. *The International Journal of Logistics Management*, *29*(1), 131–151. https://doi.org/10.1108/ijlm-11-2016-0274

World Bank (2019) https://blogs.worldbank.org/opendata/worlds-population-will-continue-grow-and-will-reach-nearly-10-billion-2050

Chapter 14

Smart Weeding Systems

Advancements and Challenges

Rupali Agarwal and Ankit Agrawal

14.1 INTRODUCTION

Agriculture is a crucial sector that feeds the world's population, but it faces several challenges, including weed management. Weeds compete with crops for resources, reducing yields and impacting food production [1]. So weed control is essential for successful crop production. Weed populations within a field tend to exhibit relative constancy from year to year, in contrast to sporadic occurrences of insect and disease outbreaks, which can yield dramatic impacts [2]. In response, farmers and vegetation managers can formulate a strategic weed management program informed by prior knowledge of anticipated weed varieties. This plan should encompass a systematically reasoned approach to weed control that emphasizes the harmonious integration of control methods with all other practices influencing the agro ecosystem. Moreover, this approach establishes a synergistic connection between weed control and the broader pest management strategy [3].

Integrated Weed Management (IWM) encompasses the utilization of diverse technological tools and a foundation of supportive knowledge in the purposeful curation, integration, and execution of effective weed control tactics. In this endeavor, careful consideration is given to the economic, ecological, and socio-logical implications of the strategies employed. Importantly, IWM operates as a fundamental component of Integrated Pest Management (IPM). The framework of IPM commonly comprises three essential elements: firstly, the application of multiple compatible pest management strategies; secondly, the maintenance of pest populations at levels that prevent economic harm; and thirdly, the conservation of environmental integrity [4]. Designing an Integrated Weed Management (IWM) system for a single crop within a single year may appear straightforward; however, the achievement of enduring success in IWM necessitates the harmonization of the farmer's attitudes, knowledge, preferences, and competencies with the most suitable tools available for each unique circumstance. The process of planning requires a comprehensive understanding of the historical weed and cropping patterns at the site, as well as a grasp of weed biology

 DOI: 10.1201/9781003676102-14

and ecology, coupled with familiarity with various weed control methodologies. Subsequently, the farmer must adroitly employ this accumulated knowledge to steer the system towards achieving robust, high-quality crop yields, all the while endeavoring to mitigate and progressively diminish the detrimental impact of weeds over time. A triumphant IWM system is one that is not only efficacious but also economically and ecologically astute. It underscores the integration of diverse control strategies with all other practices that wield influence over the ecosystem. Furthermore, it forges a cohesive bond between weed control and the broader canvas of ecosystem management [4]. Weed management practices are conventionally categorized into three tiers: prevention, control, and eradication. To attain the desired degree of weed suppression, the implementation of specific weed management techniques is imperative. While these strategies are presented as discrete methods, it is paramount to recognize that the most efficient and cost-effective weed control strategy invariably entails the synergistic utilization of multiple approaches. Each facet contributes to the overall efficacy of weed control. Neglecting or curtailing the impact of one or more components necessitates a heightened reliance on the remaining weed management practices [3]. The landscape of weed control tactics is primarily divided into six key domains: (1) scouting, (2) prevention, (3) mechanical interventions, (4) cultural practices, (5) biological control, and (6) chemical management [3]. For those intrigued by alternative perspectives, Bender's notable publication in 1994 [5] offers an exploration of weed management devoid of chemical inputs.

The aforementioned traditional weed control methods, such as manual labor and herbicide use, have limitations in terms of efficiency, cost, and environmental impact [6]. Smart weeding systems refer to advanced technologies and methods designed to automate and improve the process of removing weeds from agricultural fields or other areas [7, 8]. Traditional weeding methods often involve manual labor or the use of herbicides, which can be time-consuming, labor-intensive, and may have negative environmental impacts. Smart weeding systems aim to address these challenges by utilizing various technologies to identify and target weeds more effectively and efficiently and offer a promising solution by harnessing cutting-edge technologies to optimize weed detection and control. This chapter explores the key components, benefits, challenges, and future prospects of smart weeding systems in modern agriculture.

14.2 COMPONENTS OF SMART WEEDING SYSTEMS

Smart weeding systems are advanced technologies designed to automate and optimize the process of weed control in agricultural fields or other areas. These systems utilize a combination of sensing technologies, data analysis, and precision intervention methods to identify and remove weeds more

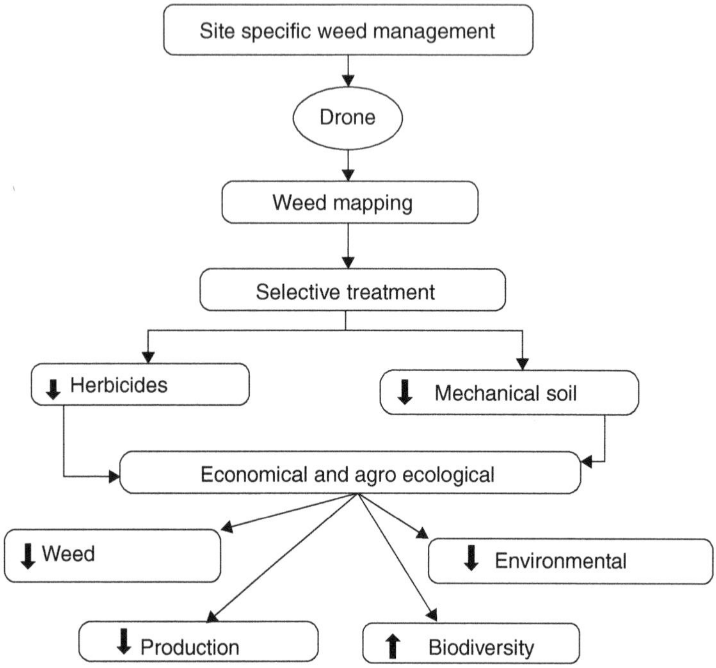

Figure 14.1 Key Components of the Smart Weeding System.

effectively and efficiently. The key components of a smart weeding system include (Figure 14.1).

14.2.1 Sensing and Imaging Technologies

Smart weeding systems rely on advanced sensors and imaging technologies to gather valuable data about the crop and weed distribution. Cameras, LiDAR, and hyperspectral sensors capture detailed information, including crop health, weed density, and growth patterns [1]. This data serves as the foundation for subsequent decision-making processes.

14.2.1.1 Cameras

High-resolution cameras capture visual information about the field, helping to identify and differentiate crops from weeds based on their appearance and growth patterns [9]. Integrated weed management, in conjunction with the utilization of Unmanned Aerial Vehicles (UAVs), enables the implementation of site-specific weed management, a highly efficient and environmentally beneficial approach. The identification of weed clusters within cultivated fields is achieved through a fusion of drone-based image capture and subsequent analysis using machine learning methodologies. Specific algorithms can be trained to guide autonomous weeding robot systems in performing weed removal, employing methods such as herbicide application or mechanical techniques. Nevertheless, a comprehensive grasp of the targeted objectives and available technologies is imperative for the rapid advancement of this domain. Within this review, we furnish an encompassing perspective on precision weed control, with a particular emphasis on the potential and pragmatic application of cutting-edge sensors accessible in the market. Substantial endeavors are requisite to attain a comprehensive understanding of weed population dynamics and their interplay with crops, facilitating the practical integration of this strategy within genuine agricultural scenarios [1].

14.2.1.2 LiDAR (Light Detection and Ranging)

LiDAR sensors use laser pulses to create precise 3D maps of the field's vegetation [10, 11]. This information aids in identifying the location and density of weeds. In [10], Light Detection and Ranging sensors were accessed to detect and locate weeds. To achieve this objective, two experiments were conducted involving simulated objects (representing weeds) placed at varying heights and diameters. The goal was to comprehend the detection capabilities of a LiDAR system. The findings indicated that the detectability of these objects at different scanning distances from the LiDAR was directly influenced by their size and alignment relative to the LiDAR device. In a subsequent trial, a wheat field was utilized as a test bed. The LiDAR system was employed to scan diverse weed species positioned at deferent elevations above the wheat canopy. This trial aimed to validate the stationary LiDAR's efficiency in identifying weeds within an actual field setting. Encouragingly, the outcomes revealed a 100 percent detection rate for weeds present in the wheat plot, with their identification primarily based on variations in height compared to the crop canopy.

14.2.1.3 Hyperspectral sensors

Hyperspectral sensors measure the reflectance of different wavelengths of light, providing detailed spectral information about the crops and weeds. This enables better weed detection and species classification [12]. Weeds are a ubiquitous presence across global farmlands. Their relentless competition for light, water and nutrients often compromises desirable plants, introduces ailments or viruses and attracts destructive insects and pests, ultimately leading to decreased yields. In response, modern weed detection technologies have emerged, aiming to enhance detection speed and precision, thereby reconciling the dual objectives of optimizing soil health and achieving effective weed control for profitable agriculture. In recent times, an array of platforms, including satellites, airplanes, Unmanned Aerial Vehicles (UAVs), and close range devices have become increasingly accessible for capturing hyperspectral imagery with varying spatial, temporal and spectral resolutions. The categorization of plants into crops and weeds based on their distinct species is pivotal for accurate weed detection.

14.2.2 Data Analysis and Machine Learning

Weeds present a significant obstacle for achieving optimal crop yield production. Although herbicides offer an effective solution, their costliness and potential for environmental harm, along with the emergence of herbicide-resistant strains, raises concerns. The application of Machine Learning (ML) models to accurately identify weeds has proven instrumental in curbing excessive herbicide usage. This research delves into the essential ML techniques employed for weed identification in corn, specifically encompassing classification and object detection. Additionally, the study examines various metrics employed to assess the efficacy of ML methodologies [13]. The subsequent sections provide a concise overview of the relevant trends in this field. A prominent facet is the utilization of digital images as primary input for most ML models, with instances of hyperspectral data adoption in select cases. Notably, support vector machines and neural networks stand out as the prevailing choices for weed identification across existing research. Evaluative measures predominantly include classification accuracy and the F1 score, serving as key indicators of ML model performance.

The large data sets generated by the sensing technologies are analyzed using machine learning algorithms and computer vision techniques. Through training on labeled data, these algorithms can accurately differentiate between crops and weeds, allowing for precise weed detection. As the system processes more data, it continually refines its weed identification capabilities, improving overall performance. The data collected from the sensing technologies are processed and analyzed using advanced data analysis techniques and machine learning algorithms. Machine learning models are trained on large datasets to recognize patterns and characteristics of weeds and crops. There are a number of key aspects of data analysis and machine learning in smart weeding systems explained below.

14.2.2.1 Feature extraction

Feature extraction refers to the process of transforming raw data into numerical features that can be processed while preserving the information in the original data set. So, the relevant features are extracted from the sensor data to distinguish between crops and weeds effectively. It yields better results than applying machine learning directly to the raw data [14,

15]. The eradication of weeds holds significant importance within the realm of agriculture. The conventional method of weed removal is not only time-intensive but also demands a considerable amount of manual labor. The focal point lies in the automated expulsion of weeds from agricultural fields. The proposed approach revolves around the utilization of deep learning techniques to identify weeds that have sprouted amongst crops. Subsequently, an automated cutting mechanism is employed to remove the detected weeds. Deep learning serves as the foundation for extracting pertinent features from agricultural images. The dataset undergoes training to facilitate the distinction between weeds and crops. Within the domain of deep learning, the Convolutional Neural Network (CNN) is harnessed. This network leverages convolutional layers coupled with Rectified Linear Unit (ReLU) functions to extricate image features. Furthermore, it incorporates max-pooling and fully connected layers with ReLU activation to discern weeds from crops. The process commences by subjecting the pre-processed image to the CNN architecture. The ensuing outcome leads to the extraction of Regions of Interest (ROI), alongside the extraction of features that contribute to the training process. Post-training, the classification phase ensues, enabling the identification of weeds. This intricate process rests upon the shoulders of a deep learning network. To enhance precision, a corpus of 100 images is utilized for training purposes [16].

14.2.2.2 Classification algorithms

In the realm of machine learning, classification algorithms leverage input training data to forecast the probability of upcoming data aligning with specific pre-established categories. Among its widespread applications, classification prominently features in tasks such as segregating emails as either 'spam' or 'non-spam'. In essence, classification embodies a variant of 'pattern recognition,' where the algorithms scrutinize training data to identify recurring patterns (such as analogous words, sentiments, numerical sequences, and the like). These recognized patterns are subsequently utilized to categorize forthcoming datasets. Machine learning models, such as Support Vector Machines (SVM), decision trees, or deep neural networks, are used for weed classification and identification [17].

14.3 WEED DETECTION AND IDENTIFICATION

The traditional method employed to eliminate weeds within a crop plantation involves the widespread application of herbicides across the entire area. Unfortunately, this approach leads to the pollution of the food crops and a reduction in yield due to unintended damage to both crop plants and weeds. Consequently, there arises a pressing necessity for the development and implementation of an intelligent or smart weed control system. Once the data analysis is complete, the smart weeding system can identify and classify weeds present in the agricultural field [8]. The system may recognize weeds based on various features, such as leaf shape, color, size, or growth patterns. Some systems may also consider other environmental factors, like soil moisture or temperature, to enhance weed detection accuracy. The combination of sensing technologies and data analysis enables smart weeding systems to accurately detect and identify weeds among crops. These systems can determine weed species, density, and locations, which are essential for targeted intervention.

14.3.1 Techniques of Smart Weeding

Once the weeds are identified, smart weeding systems employ precision spraying [18] or mechanical weeding techniques to remove the weeds. These targeted interventions help minimize the use of herbicides and reduce environmental impacts. In response to the aspiration to curtail chemical usage in agriculture, farmers seek alternative methods for weed management. Unfortunately, conventional mechanical cultivation lacks the capacity for selective weed removal, and certain weed scenarios lack appropriate selective herbicides. Given the cost-intensive nature of manual labor, the viability of an automated weed control system becomes a plausible consideration. Such a robotic system holds the potential to diminish or entirely eradicate the dependence on chemical interventions. Numerous endeavors have been undertaken to devise effective algorithms for the recognition and categorization of objects. Presently, ongoing research is dedicated to the advancement of novel machine vision algorithms aimed at the automated recognition and classification of diverse object groups.

14.3.1.1 Chemical weeding by precision spraying

Precision spraying systems use technologies like individual nozzle control or robotic sprayers to apply herbicides only to the identified weeds, reducing chemical usage and minimizing off-target effects. For example, in [8], utilizing an image processing algorithm, photographs of the crop plantation rows are systematically captured at regular intervals. Upon the detection of weeds within these images, a precise and targeted application of herbicide exclusively onto the weed-infested areas ensues. The algorithm primarily relies on an erosion and dilation methodology for weed identification. The process involves converting the color image into a binary representation, with emphasis on isolating the green components. The analysis quantifies the presence of white pixels within the defined region of interest. Subsequently, regions exhibiting a white pixel count surpassing the predefined threshold are identified as weed-infested areas. The herbicidal solution is housed within a container equipped with water pump motors linked to specialized spray nozzles. Once the weeds are successfully identified, a signal is dispatched from a Raspberry-Pi module to a motor driver IC, which exercises control over the water pump motors. This orchestrated mechanism triggers the precise release of chemicals, effectively targeting and eradicating the identified weeds while safeguarding the surrounding crop plants.

Machine vision applied to selective weeding or targeted herbicide spraying heavily relies on the system's capability to meticulously analyze weed images and process the derived insights. This informed analysis precedes the execution of a discerned control strategy. Tailoring the weed control requires distinct herbicide formulations for different weed types, thereby necessitating accurate identification and subsequent classification of weeds. In the work of Juraiza Ishak et al. [14], weed images were categorized into either broad or narrow weed types. A core challenge in weed image recognition using planar curve analysis is the accurate detection of curves. In the work of Siddiqi et al. [19], a self-regulating spray control system is facilitated through the creation of an algorithm. This algorithm is founded upon a segmentation approach, commencing with erosion and culminating in dilation. Its functionality extends to both the identification and classification of weeds. After the results of this research, Amir et al. [15] developed a new weed detection and classification method which could be used in autonomous weed control robots. Veeragandham and Santi [20] investigated whether among the array of machine learning algorithms, the Convolutional Neural Network (CNN) emerges as a formidable choice for the precise classification of both crops and weeds. Its proficiency is particularly pronounced in challenging scenarios, encompassing variables such as diverse lighting conditions, plant species differentiation, instances of overlapping crops and weeds, and the development of automated patch sprayers.

14.3.1.2 Mechanical weeding

Some smart weeding systems use autonomous robots or specialized machinery equipped with tools to physically remove weeds without damaging crops. The emergence of herbicide-resistant weeds has significantly amplified the significance and potential of mechanical weed control [21]. In various regions across the globe, the efficacy of mechanical weed control measures has been subject to reevaluation. Despite being one of the oldest techniques in the weed control arsenal, tillage, employed for soil preparation before planting, remains the favored approach for many farmers.

Weeds impose substantial losses on crops, prompting the implementation of effective control measures. In conventional management systems, manual, mechanical, and chemical methods are commonly employed for weed control [22]. Specifically, it highlights the prospects presented by Rhizobacteria (soil-borne bacteria which live on or around the roots of plants in an area

Figure 14.2 Mechanical Weed Control Methods.

referred to as the rhizosphere) as a promising avenue for innovative weed management strategies. Various types of mechanical weed control methods are given in Figure 14.2.

Tillage: Tillage serves to eliminate weeds from the soil, leading to their demise. However, it can inadvertently harm plants by causing injuries or pruning of their roots and stems, thereby diminishing their ability to compete and regenerate. Furthermore, tillage has the effect of burying weeds beneath the soil surface. In addition to its weed control function, tillage significantly influences the overall physical state of the soil. Basic plowing, for instance, aids in the burial of annual, biennial, and numerous perennial weeds. The primary tools utilized for tillage include implements such as the plow and harrow, among others.

Hoeing: For centuries, the hoe has remained the quintessential and extensively employed tool for weeding. The iconic image of 'The man with the hoe,' emblematic of weed management throughout history, is gradually being supplanted by 'The man with the sprayer.' Nevertheless, the hoe retains its practical utility as a means to achieve cost-effective and efficient outcomes. Hoeing, especially adept at eradicating annual and biennial weeds, remains a potent method for complete weed removal.

Hand weeding: Hand weeding involves the manual extraction or uprooting of weeds using one's hands or a tool known as a 'khurpi,' which bears a resemblance to a sickle. This technique likely stands as the most ancient means of weed management, enduring as a viable and effective approach for eradicating weeds across both cultivated and uncultivated terrains.

Digging: Digging proves highly advantageous when contending with perennial weeds, as it enables the extraction of subterranean propagating elements from the lower soil strata. This method surpasses the efficacy of mere hand pulling. The act of digging is complemented by the manual removal of the plant. However, if proper digging is neglected, mere hand pulling might merely eliminate surface growth, permitting the resurgence of vegetation from residual plant components.

Sickling and mowing: Manual sickling entails the use of a sickle by hand to trim the upper foliage of weeds, halting seed generation and depriving subterranean components of sustenance. This technique is similarly employed to curtail the expansion of fully developed perennial weeds. Mowing, conversely, is a mechanized operation frequently conducted along roadways and across lawns. Its aims align closely with those of sickling.

Burning: Burning or employing fire frequently serves as a cost-effective and feasible strategy for weed management. This approach is utilized to: Eliminate amassed plant matter, Exterminate the dried upper portions of fully grown weeds, Eradicate actively growing green weeds, and annihilate latent weed seeds buried in the soil. The utilization of fire extends to the incineration of crop remnants in various crops such as cotton, sugarcane, potatoes, corn, soybean, sorghum, castor, and others.

Flooding: Occasionally, flooding is employed as a method of weed management in rice fields, given rice's ability to thrive in flooded environments. The inundation process diminishes the availability of oxygen necessary for plant growth, leading to the demise of unwanted vegetation. The effectiveness of flooding as a treatment hinges on the thorough submersion of weeds for extended durations.

Mulching: Mulching involves the practice of shielding the soil using organic or non-reactive substances like straw, sawdust, bark dust, and assorted plant components, ranging in size. Alternatively, materials such as paper, plastic sheets, or polythene can also be used. Mulching serves as a viable technique on a limited scale within certain regions to curb the proliferation of weeds.

14.4 BENEFITS AND IMPACT

Smart weeding systems offer numerous benefits that significantly enhance weed control in agriculture and other applications. These benefits stem from the integration of advanced technologies like artificial intelligence, sensing systems, and autonomous operation. Here are some key advantages of smart weeding systems:

1 Increased efficiency: Smart weeding systems can accurately detect and target weeds with high precision. By selectively applying interventions only where necessary, such as precision spraying or mechanical weeding, these systems minimize unnecessary herbicide use and reduce labor requirements. This increased efficiency leads to optimized weed control and resource utilization.
2 Improved crop yields: By minimizing weed competition with crops, smart weeding systems help improve overall crop yields. Weeds can deprive crops of essential nutrients, water, and sunlight. Effective weed control ensures crops receive the resources they need for healthy growth and higher yields.
3 Environmental sustainability: Smart weeding systems contribute to more sustainable agricultural practices. They reduce the reliance on chemical herbicides by applying them only to targeted weeds, thus minimizing environmental contamination and potential harm to

non-target species. Additionally, the use of mechanical weeding and organic practices further promotes eco-friendly farming.

4 Reduced herbicide resistance: Conventional herbicide applications often lead to the development of herbicide-resistant weed species, which poses a significant challenge to weed management. Smart weeding systems, by using localized and targeted herbicide applications, can help delay or mitigate the development of herbicide resistance, preserving the effectiveness of herbicides.
5 Labor savings: Automation and autonomous operation in smart weeding systems reduces the need for manual labor in weed control. Farmers can focus on other essential tasks while the smart weeding system handles weed detection and removal, leading to labor savings and potentially freeing up time for other activities.
6 Adaptability to changing conditions: Smart weeding systems are designed to adapt to changing environmental conditions, weed types, and growth stages. The integration of AI allows the system to learn from real-time data and adjust strategies accordingly, ensuring optimal weed control even as conditions evolve.
7 Precision weed identification: With advanced sensing technologies and AI algorithms, smart weeding systems can accurately identify different weed species. This precision helps target the most problematic weeds effectively, leading to more successful weed management.
8 Minimization of soil disturbance: Mechanical weeding in smart weeding systems can be tailored to minimize soil disturbance compared to conventional plowing or tilling. This helps retain soil structure and reduces the risk of soil erosion and nutrient loss.
9 Enhanced data-driven insights: Smart weeding systems generate vast amounts of data on weed distribution and growth patterns. These data insights can be valuable for farmers to make informed decisions on weed management strategies, optimize crop rotation plans, and improve overall farm planning.
10 Support for precision agriculture: Smart weeding systems complement other precision agriculture technologies by providing localized and real-time information on weed infestations. This integration helps farmers make data-driven decisions that can improve the overall sustainability and productivity of their operations.

14.5 CHALLENGES AND FUTURE PROSPECTS

1 Advancements in AI and machine learning: As AI and machine learning continue to evolve, smart weeding systems will benefit from more sophisticated algorithms for weed detection, identification, and decision-making. This will lead to higher accuracy, adaptability, and efficiency.

2 Miniaturization and cost reduction of sensors: Advancements in sensor technology can lead to smaller more affordable, and highly capable sensors for weed detection. This can reduce the overall cost of smart weeding systems and enhance their accessibility to a wider range of farmers.
3 Edge computing and real-time processing: Edge computing can empower smart weeding systems to process data locally and make real-time decisions without relying heavily on cloud-based infrastructure. This can improve response times and reduce the reliance on stable internet connections.
4 Collaborative multi-robot systems: Future smart weeding systems may involve collaborative multi-robot systems, where multiple autonomous robots work together to cover larger areas and enhance weed control efficiency.
5 Smart farming integration: Integration with other smart farming technologies, such as precision agriculture, the Internet of Things (IoT) devices, and drone-based monitoring, can lead to more comprehensive and data-driven farm management strategies.
6 Energy harvesting and sustainability: Research into energy harvesting technologies can help make smart weeding systems more sustainable by powering them through renewable energy sources, reducing reliance on traditional power supplies.
7 Human-AI collaboration: Future smart weeding systems may embrace human-AI collaboration, where farmers work alongside AI-powered robots, leveraging human expertise and AI capabilities for optimal weed management.

14.6 CONCLUSION

Smart weeding systems represent a transformative approach to weed management in modern agriculture. By combining advanced sensing technologies, AI-driven algorithms, and targeted intervention methods, these systems offer an efficient, sustainable, and labor-saving solution for weed control. As the technology continues to develop and become more accessible, smart weeding systems have the potential to revolutionize farming practices, supporting global food security and environmental sustainability. In conclusion, the evolution of smart weeding systems represents a significant leap forward in modern agriculture and weed management practices. These systems, powered by advanced technologies such as artificial intelligence, computer vision, and automation, offer a host of benefits and solutions to the challenges posed by conventional weed control methods.

Smart weeding systems excel in precision and efficiency, allowing for the selective targeting of weeds while minimizing the use of herbicides and reducing labor costs. They contribute to increased crop yields, environmental sustainability, and the preservation of soil health. By leveraging real-time data and AI-driven decision-making, these systems adapt to changing conditions and offer valuable insights for farmers.

Furthermore, the integration of smart weeding systems with other precision agriculture technologies enables a holistic approach to farm management, promoting sustainability and resource optimization. However, it is essential to address challenges such as weed identification accuracy, energy efficiency, and safety to fully unlock the potential of these systems.

The future prospects of smart weeding systems are promising. Advancements in AI, sensor technologies, and robotics are poised to make these systems even more effective and accessible. Collaborative multi-robot systems, energy harvesting solutions, and human-AI partnerships are likely to shape the next generation of smart weeding technologies.

REFERENCES

1. C. Valerio, S. Fabrizio, M. Albino, E. Marco, and C. Mariano. Drone and sensor technology for sustainable weed management: A review. *Chemical and Biological Technologies in Agriculture*, vol 8, 8–18, 2021.
2. Lata Rani, Komal Thapa, Neha Kanojia, Neelam Sharma, Sukhbir Singh, Ajmer Singh Grewal, Arun Lal Srivastav, and Jyotsna Kaushal. An extensive review on the consequences of chemical pesticides on human health and environment. *Journal of Cleaner Production*, 283:124657, 2021.
3. Thomas J. Monaco, Floyd M. Ashton, and Stephen C. Weller. *Weed Science: Principles and Practices*. John Wiley, 2002.
4. J. M. Lish, R. H. Callihan Thill, D. C. and E. J. Bechinski. Integrated weed management component of integrated pest management: A critical review. *Weed Technology*, 5:648–656, October 1991.
5. J. Bender. *Future Harvest: Pesticide-Free Farming*. University of Nebraska Press, 1994.
6. Mohd Marzuki Mustafa, Kamarul Hawari Ghazali and Aini Hussain. Machine vision system for automatic weeding strategy using image processing technique. *American-Eurasian Journal of Agricultural Environmental Sciences*, 3:451–458, 2008.
7. Ajit G. Deshmukh and V.A. Kulkarni. Advanced robotic weeding system. *ITSI Transactions on Electrical and Electronics Engineering (ITSI-TEEE)*, 1 :51–56, 2013.
8. R. Aravind, M. Daman, and B. S. Kariyappa. Design and development of automatic weed detection and smart herbicide sprayer robot. In *2015 IEEE Recent Advances in Intelligent Computational Systems (RAICS)*, pp. 257–261, Dec 2015.
9. Francisca Lpez-Granados, Montse Jurado-Expsito, Jose M. Pea-Barragn, and Luis Garca-Torres. Using remote sensing for identification of late-season grass weed patches in wheat. *Weed Science*, 54(2):346353, 2006.
10. Nooshin Shahbazi, Michael B. Ashworth, J. Nikolaus Callow, Ajmal Mian, Hugh J. Beckie, Stuart Speidel, Elliot Nicholls, and Ken C. Flower.

Assessing the capability and potential of lidar for weed detection. *Sensors*, 21(7):23–28, Mar 2021.

11. Dionisio Andjar, Victor Rueda-Ayala, Hugo Moreno, J. R. Rosell, Alexandre Escol, Constantino Valero, Roland Gerhards, and Hans W. Griepentrog. Discriminating crop, weeds and soil surface with a terrestrial lidar sensor. *Sensors*, 13:14662–14675, Nov 2013.
12. Muhammad Huzaifah, Mohd Roslim, Abdul Shukor Juraimi, Nisfariza Mohd Noor, Wan Fazilah Fazlil, Ilahi Nursyazyla Sulaiman, and Nursyazyla Sulaiman. The application of hyperspectral remote sensing imagery (hrsi) for weed detection analysis in rice fields: A review. *Applied Sciences*, 12(5), pp. 1–19, Mar 2022.
13. Akhil Venkataraju, Dharanidharan Arumugam, Calvin Stepan, Ravi Kiran, and Thomas Peters. A review of machine learning techniques for identifying weeds in corn. *Smart Agricultural Technology*, 3:100102, 2023.
14. Asnor Juraiza Ishak, Siti Salasiah Mokri, Mohd Marzuki Mustafa, and Aini Hus- Sain. Weed detection utilizing quadratic polynomial and roi techniques. In *2007 5th Student Conference on Research and Development*, pp. 1–5, Dec 2007, IEEE.
15. H. Amir, B. Kargar, and Ali M. Shirzadifar. Automatic weed detection system and smart herbicide sprayer robot for corn fields. In *2013 First RSI/ISM International Conference on Robotics and Mechatronics (ICRoM)*, pp. 468–473, Feb 2013, IEEE.
16. F. Joseph, B. L. Velammal, S. Manoruthra, and V. Kalaivani. Automated weed removal system using convolutional neural network. *International Journal of Scientific and Technology Research*, 9:1212–1214, March 2020.
17. Giuseppe Bonaccorso. *Machine Learning Algorithms*. Packt, 2017.
18. Gaurav Sethia, Harish Kumar S. Guragol, Swati Sandhya, J. Shruthi, and N. Rashmi. Automated computer vision based weed removal bot. In *2020 IEEE International Conference on Electronics, Computing and Communication Technologies (CONECCT)*, pp. 1–6, July 2020, IEEE.
19. Muhammad Hameed Siddiqi, Irshad Ahmad, and Suziah Bt Sulaiman. Weed recognition based on erosion and dilation segmentation algorithm. In *2009 International Conference on Education Technology and Computer*, pp. 224–228, April 2009, IEEE.
20. Syamasudha Veeragandham, and H Santhi. A detailed review on challenges and imperatives of various CNN algorithms in weed detection. In *2021 International Conference on Artificial Intelligence and Smart Systems (ICAIS)*, pp. 1068–1073, March 2021, IEEE.
21. Khawar Jabran, and Bhagirath S. Chauhan Hobo- ken, NJ 07030. *Non-Chemical Weed Control*. Elsevier, 2018.
22. Tasawar Abbas, Zahir A. Zahir, Muhammad Naveed, and Robert J. Kremer. Limitations of existing weed control practices necessitate development of alternative techniques based on biological approaches. *Advances in Agronomy*, 147:239–280, 2018.

Index

For Product Safety Concerns and Information please contact our EU representative GPSR@taylorandfrancis.com
Taylor & Francis Verlag GmbH, Kaufingerstraße 24, 80331 München, Germany

www.ingramcontent.com/pod-product-compliance
Ingram Content Group UK Ltd.
Pitfield, Milton Keynes, MK11 3LW, UK
UKHW022220190926
475702UK00003B/467
9781032565873